Y0-CIP-778

Lumped Systems

Holt, Rinehart and Winston Series
in Electrical Engineering, Electronics, and Systems

Other Books in the Series:

George R. Cooper and Clare D. McGillem, *Methods of Signal and System Analysis*
Samuel Seely, *Electronic Circuits*
Mohammed S. Ghausi and John J. Kelly, *Introduction to Distributed-Parameter Networks: With Application to Integrated Circuits*

Lumped Systems

BENJAMIN J. LEON

Professor of Electrical Engineering
Purdue University

HOLT, RINEHART AND WINSTON, INC.
New York Chicago San Francisco Atlanta
Dallas Montreal Toronto London

Library of Congress Catalog Card Number: 68-28179
2723104
Printed in the United States of America
1 2 3 4 5 6 7 8 9

Preface

This book was prepared as a text for a dual level (senior-graduate) cource in engineering system theory. In the author's opinion, everyone who gets a graduate degree in electrical engineering should have a knowledge of the theory of lumped systems that is one level more advanced than that of most undergraduate curricula. This book is a text for a course that provides this knowledge.

The text is not intended as a complete tome with every detail. Special cases are usually discussed only briefly with the details left to the problems that are included at the end of each chapter. Furthermore, many well known facts that are easily derived from material that students generally learn as undergraduates are also included in the problems. For the well trained and well motivated student, the book is suitable for self study. For most students at this level, including those whose primary interests are not in the systems areas of electrical engineering, the guidance of a professor will be required.

The background assumed for the student is the standard undergraduate curriculum in electrical engineering. He should be familiar with matrix manipulation and the algebra of complex numbers. Full courses in linear algebra and analytic function theory are not required. The student should have learned to write circuit differential equations from a network graph using the concept of a tree. He should have had some practice in frequency domain analysis and have become familiar with the description of lumped, linear, and time-invariant systems via poles and zeros. Further, the student is assumed to have been familiar with differential equations of mechanical systems and with electrical mechanical analogs. In all these areas a real depth of understanding is not assumed at the beginning, but by the end of the course it should have been acquired.

The book is divided into three parts. The first part, which consists of five chapters, is basically an analysis of lumped systems using state variables as the underlying technique. For the student whose undergraduate course was presented from the state variables point of view, much of the material will be review. For those with a more traditional background based exclusively on operational calculus, Part one is the presentation of a new method for performing old tricks. The relation between state variables methods and operational methods are brought out. By the end of the course the student should see the justification for learning both.

Part two of the text contains three chapters, two of which deal with some important properties of lumped, linear, and time-invariant systems. The short Chapter 7 is devoted to some properties of transforms – properties required for an understanding of systems and usually not included in undergraduate courses. The system properties discussed include the time domain concepts of causality, passivity, and stability, and the frequency domain concept of positive realness. In addition, frequency domain relations between magnitude and phase shift of transfer functions are considered.

The third part of the text consists of two chapters on nonlinear systems. The first of these introduces basic concepts such as phase space trajectories and stability in the sense of Lyapunov. This chapter plus the preceding eight completes the development of "what every MSEE should know." The final chapter is an introduction to some more advanced topics in nonlinear systems – research topics of particular interest to the author. When the course is taught by someone else, the final topic, time permitting, will most probably be different. One possibility is linear system theory via the scattering matrix – a topic that is the logical pièce de résistance to follow Part two of the book. Another such advanced topic that logically follows Part one is optimum control. Rather than do a mediocre job on an introduction to a research area other than his own, the author leaves the complete freedom of presenting the final topic to the professor.

As a final point about the place of this text in engineering education we should consider where the student goes from here. The student whose interests are primarily in electrophysics goes back to his specialty. The student in the systems areas of electrical engineering must develop a more advanced level of mathematical maturity before he can proceed. The first nine chapters of the book develop techniques without utilizing any very subtle mathematical ideas. In Chapter 10 the basic mathematical facts used are still elementary, but their application is more subtle. To proceed beyond Chapter 10, or on a similar level of introduction to a research area in system theory, requires more and more mathematical maturity. The serious student in the systems are should develop his mathematical skills before proceeding.

During the development of the material for this text there were several topics that required further research, because they were not adequately developed in the existing engineering literature. The research efforts of the author were supported during this period by the National Science Foundation. Without this support, Chapters 5, 8, 9, and 10 would not be as complete as they appear now. The thorough reading and commentary by Professor M. Athans contributed greatly to the development of the final manuscript. Others who made major contributions include the author's research colleagues at Purdue: Professors D. R. Anderson, L. O. Chua, Y. L. Kuo, and P. M. Lin.

Lafayette, Indiana BENJAMIN J. LEON

August 1968

Contents

Preface v

PART I—ANALYSIS OF LUMPED, LINEAR SYSTEMS 1

Chapter 1—Preliminaries 3
1-1 Introduction
1-2 Terminology
1-3 A Simple Example

Chapter 2—Normal Form Equations from System Diagrams 14
2-1 Introduction
2-2 The Electric Circuit Diagram
2-3 The Mechanical Circuit Diagram
2-4 High-Order Equations and Analog Computer Diagrams

Chapter 3—Solution of Normal Form Equations by Variation of Parameters 34
3-1 Fixed-Coefficient Systems

3-2 Some Remarks on Time-Variant Systems
3-3 Some Simpler Computational Techniques

Chapter 4 — General Linear Systems Described by the Convolution Integral 53
4-1 Properties of Convolutions
4-2 Block Diagram System Representation
4-3 The Total Response — Zero State Plus Zero Input

Chapter 5 — Discrete Time Signal Processing 87
5-1 Time-Domain Input-Output Relations for Lumped Systems
5-2 Summation of Discrete Convolutions — The *Z*- Transform
5-3 *Z*-Transform Analysis of Lumped Systems
5-4 Linear Discrete Time Systems

PART II — SOME CLASSIFICATIONS FOR LUMPED, LINEAR, TIME-INVARIANT SYSTEMS 109

Chapter 6 — Systems with Positive-Semidefinite Energy Functions 111
6-1 Preliminaries
6-2 Resistive Systems
6-3 Lossless Systems
6-4 The General System of Equation (6-1)

Chapter 7 — Properties of Laplace-Fourier Transforms 134
7-1 Basic Properties of Two-Sided Transforms
7-2 Even and Odd Functions

Chapter 8 — Properties of Lumped Systems Characterized by *PR* Matrices 143
8-1 The Poles and Zeros of *PR* Matrix Elements
8-2 Causality of *PR* Systems and Relations between Real and Imaginary Parts
8-3 Stability for *PR* Systems
8-4 Passivity for *PR* Systems

PART III — SOME PROPERTIES OF AND TECHNIQUES APPLICABLE TO NONLINEAR SYSTEMS 167

Chapter 9—The Use of State Variable and Energy Function Concepts 169
9-1 Formulation
9-2 Second-Order Systems—Trajectories in the Phase Plane
9-3 Some Additional Stability Notions

Chapter 10—Iteration of Nonlinear Systems Problems 186
10-1 The State at t_0
10-2 Bounds on the Response
10-3 Stability Using Both Integral and Differential Equations

Bibliography 215

Index 217

Lumped Systems

PART I

Analysis of Lumped, Linear Systems

CHAPTER 1

Preliminaries

1-1 Introduction

In engineering system analysis, one starts with a given physical system and constructs a diagram model. This model in turn is readily described by equations. From the equations the construction of a solution is a purely mathematical process. This solution of the equations is called the *response of the model.* If the model is a good one, the response of the physical system is essentially the same as the response of the model. In design the engineer generates a diagram model whose response has useful engineering properties. This diagram should be one that the engineer knows to be a good model of a physical system that can be constructed. In this text the analysis is based on the diagram. The connection between the diagram and real objects is not considered.

In this text the first problem in analysis is the step from the diagram to a set of equations in a convenient form. When the system equations are differential equations, the most convenient form for the general case is a set of first-order differential equations. The advantages of this form of equations, advantages that will be more apparent after a discussion of the solutions, are the relations between important physical properties of the system and easily computed mathematical quantities. For any particular system with a particular response to be computed, there may well be a fast method of analysis that gets from the system diagram to the desired quantity without going through the complete set of system equations. Such methods are not discussed exhaustively herein, although some short cuts that have fairly wide applicability are introduced. Other short cuts are introduced through the problems.

From the system equations, the next step is the construction of a solution. Of the many methods for solving differential equations, the method of variation of parameters gives the most generality with the least mathematical complexity. Furthermore, the steps of this method bring out many relations between important physical concepts and well-known mathematical techniques. Once the equation solution—or system response—has been found, the relation between the general method and various short cuts, which are normally presented without proof in undergraduate circuits and systems texts, are apparent.

Since lumped, linear, time-invariant systems have a forced response in the form of a convolution integral, this mathematical operation is discussed in some detail. With the convolution background, the block-diagram model of any such system is easily discussed in terms of first-order differential equations. The construction of a set of first-order equations from a block diagram of interconnected transfer functions completes the analysis problem for lumped, linear, time-invariant systems.

The analysis of discrete systems is presented in Chapter 5. Although a discrete time system is not lumped according to the definition given below, the similarities between the system equations and solution methods for these systems and those for lumped systems justifies inclusion. The similarities and differences between true discrete-time systems, and lumped systems processing discrete time signals are presented in some detail.

1-2 Terminology

Before proceeding further, the terms introduced above and some other familiar terms that are used in this chapter need to be defined.

The word *system* can mean almost anything. Various authors have tried to give a precise mathematical definition of a system. In this text the definition is applied more as a dictionary definition. In addition to the word "system," words such as "physical," "signal," "component," and so on, will be defined as in a dictionary. The reader should not expect to be able to apply a precise mathematical criterion to each use of the word. Words including "linear," "time-invariant," "stable," and so on, are precisely defined and the uses of the words can be checked mathematically.

The following is not an exhaustive list of the terms used in the text; others will be defined as needed. The word system should be defined first; but since its definition requires others, the ordering is made so each definition requires only words already defined.

Physical: An adjective that refers to any quantity that is associated with the basic quantities of physics – length, mass, time, charge.

Mapping: A rule, formula, or set of formulas for going from one set of objects to another.

Function: A mapping from one set (called the domain) to another set (called the range) that assigns only one member of the range to each member of the domain. The word function will also be used to mean a member of the range of the function. The precise meaning should be clear from the context. In this text there is no such thing as a multiple-valued function.

Signal: A physical function of time, for example, voltage, current, force, velocity, temperature, pressure.

System: A signal processor (a mapping from a set of input signals to a set of output signals). By tying systems and signals to the physical world, the generality is somewhat limited; however, the interpretations and results are still general enough for electrical engineering purposes.

Element or Component: A system that is part of another system.

Lumped System: A system that can be decomposed into a finite number of components, each with a finite number of inputs and outputs and such that the values of the outputs at every time are functions of the inputs, their derivatives and integrals, at the same time. The rules for interconnecting these components are additive. As a consequence of this definition, lumped systems are characterized by ordinary differential equations.[1] Since the velocity of light and the velocity of sound are finite, all real physical objects with inputs and outputs at different places are not lumped. Nevertheless, the lumped-system diagram can describe a physical system to a good approximation.

---

[1] Schwartz and Friedland (Reference 22) define a lumped system as a system characterized by ordinary differential or difference equations. In allowing difference equations they include sections of distortionless transmission line. By using a definition based on the diagram rather than the system equations such physically distributed elements are avoided.

State of a System at Time t_0: The minimum amount of information needed about the system at time t_0, in addition to the system diagram (or system equations) and the system inputs between t_0 and t_1, so that all outputs at any time t_1 can be determined. An example is the *RLC* circuit wherein the energy storage in the capacitor and the inductor at t_0 is enough to determine the total response at any time t_1 if the excitation is known between t_0 and t_1.

Linear System: A system whose input-output mapping over the interval (t_0, t_1) is linear when the effect on the output of the state at t_0 can be considered as a separate additive term. More specifically, a linear system is defined by three computations (measurements) and a test as follows:

1. Compute the set of output signals $\{S_{oo}(t)\}$ over the interval (t_0, t_1) when the input signals $\{S_i(t)\}$ are all zero on that interval.

2. Choose input signals $\{S_{i1}(t)\}$ and compute the output signals $\{S_{o1}(t)\}$.

3. Choose input signals $\{S_{i2}(t)\}$ and compute the output signals $\{S_{o2}(t)\}$.

4. For any constants A_1 and A_2, the system outputs must be $A_1[\{S_{o1}(t)\} - \{S_{oo}(t)\}] + A_2[\{S_{o2}(t)\} - \{S_{oo}(t)\}] + \{S_{oo}(t)\}$ when the inputs are $A_1\{S_{i1}(t)\} + A_2\{S_{i2}(t)\}$.

The braces around the signal designation indicate that there are, in general, several input signals and several output signals. The test in step 4 must be valid; independent of the choice of $\{S_{i1}(t)\}$, $\{S_{i2}(t)\}$, A_1, and A_2. Of course, for all four computations, the state at t_0 must be the same.

In Chapter 4 it is shown that the response due to the state at t_0 is linearly related to a set of numbers that determine the state, and that these numbers can be related to a set of input signals in a system whose state at t_0 is zero.

Time-Invariant System: A system for which a time-shift of the input signal induces the same time-shift on that part of the output signal due to the input. In the symbols used in the definition of linearity, the time-invariance definition is: The input-output relation

$$\{S_i(t)\} \rightarrow \{S_o(t)\}$$

implies

$$\{S_i(t+a)\} \rightarrow \{S_o(t+a)\} - \{S_{oo}(t+a)\} + \{S_{oo}(t)\}$$

for all real a. The term S_{oo} is the zero-input response of step 1 in the definition of linear system.

1-3 A Simple Example

Before discussing the general solution of lumped, linear systems, let us review a simple example. The simplest system that can be used to demonstrate the general solution method is the *RLC* circuit, shown in Figure 1-1.

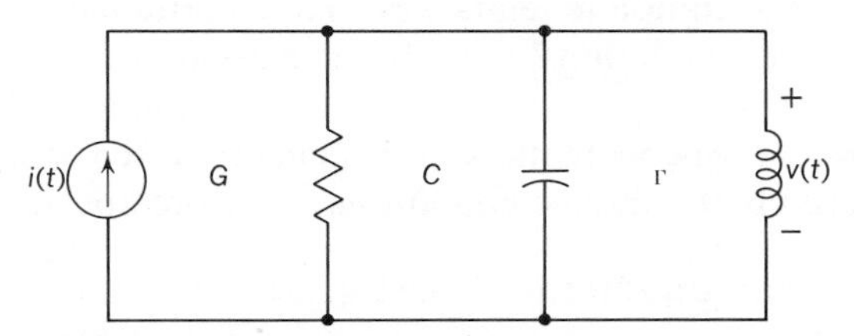

Figure 1-1 A simple second-order system.

For this circuit, Kirchhoff's current law gives

$$i(t) = G\,v(t) + C\,\dot{v}(t) + \Gamma \int v\,dt \tag{1-1}$$

where the dot over the symbol means "derivative with respect to time," and the integral is left indefinite for the present. We could now proceed with the method of variation of parameters or through the use of Laplace transforms and solve Equation (1-1). Because of the integral in (1-1) there is some uncertainty as to how to apply the method of variation of parameters; for Laplace transforms the constant of integration is set by the inductor current at $t = 0$. The question of initial conditions and the constant of integration is delayed if we use the flux linkage λ as a second variable—the voltage v being the first variable. Now Equation (1-1) reduces to two coupled first-order differential equations

$$\begin{aligned} i(t) &= G\,v(t) + C\,\dot{v}(t) + \Gamma\,\lambda(t) \\ v(t) &= \dot{\lambda}(t) \end{aligned} \tag{1-2}$$

The second of Equations (1-2) is Faraday's law. For reasons that will become apparent in the general case, it is convenient to rewrite Equation (1-2) in the so-called *normal form*

$$\begin{aligned} \dot{v}(t) &= -\frac{G}{C}\,v(t) - \frac{\Gamma}{C}\,\lambda(t) + \frac{1}{C}\,i(t) \\ \dot{\lambda}(t) &= v(t) \end{aligned} \tag{1-3}$$

We recall (see, for example, Reference 11) the procedure for the method of variation of parameters as follows:

1. Consider the homogeneous part of the differential equation and assume a solution in the form Ae^{pt}. This form satisfies the equation

for any A and for certain values of p. These p values are the roots of a polynomial.

2. Assume that the complete equation has the same basic functional form as that of the solution to the homogeneous equation, but with the free parameter A as a function of t. Thus, the assumed form is $A(t)\,e^{p_i t}$ with the p_i determined in step 1. The result is a particular solution to the complete equation. This particular solution is not unique because a solution to the homogeneous equation can always be added.

3. Determine the constants still remaining after step 2 from the initial conditions (boundary conditions).

The homogeneous equation. For Equation (1-3) we start with the homogeneous equations

$$\dot{x}(t) = -\frac{G}{C}\,x(t) - \frac{\Gamma}{C}\,y(t)$$
$$\dot{y}(t) = x(t) \tag{1-4}$$

Here we use the symbols x and y instead of v and λ since the solutions to Equations (1-4) do not satisfy (1-3) unless $i(t)$ is zero. For step 1 we assume both x and y to have the form Ae^{pt}. For the present there is no reason to assume that the same constants A and p apply to both x and y. Thus we try

$$x(t) = Ae^{pt};\ y(t) = Be^{qt}$$

Substituting into (1-4) gives

$$Ape^{pt} = -\frac{G}{C}\,Ae^{pt} - \frac{\Gamma}{C}\,Be^{qt}$$
$$Bqe^{qt} = Ae^{pt} \tag{1-5}$$

The relation between p and q is readily determined from the second of Equations (1-5). This can be rewritten as

$$e^{(p-q)t} = \frac{Bq}{A}$$

Since B, q, and A are all constants, this equation is true if and only if $p = q$. Thus, (1-5) can be written

$$Ape^{pt} = \left(-\frac{G}{C}\,A - \frac{\Gamma}{C}\,B\right)\,e^{pt}$$
$$Bpe^{pt} = Ae^{pt} \tag{1-6}$$

Cancelling the e^{pt} from both sides and regrouping yields the two homogeneous algebraic equations for A and B

$$\left(-\frac{G}{C}-p\right)A-\frac{\Gamma}{C}B=0$$

$$A-pB=0 \tag{1-7}$$

Cramer's rule applied to Equation (1-7) shows that either both A and B are zero, or the determinant of the coefficients is zero, and the two equations are dependent. In the first instance the solution is the trivial one, that is, $x=y=0$. The second case gives the interesting result. Thus, p must be selected so that the determinant

$$\begin{vmatrix} -\frac{G}{C}-p & -\frac{\Gamma}{C} \\ 1 & -p \end{vmatrix}=0$$

or

$$p^2+\frac{G}{C}p+\frac{\Gamma}{C}=0 \tag{1-8}$$

This polynomial, called the *characteristic polynomial*, has two roots

$$p_1=-\frac{G}{2C}+\sqrt{\left(\frac{G}{2C}\right)^2-\frac{\Gamma}{C}}$$

$$p_2=-\frac{G}{2C}-\sqrt{\left(\frac{G}{2C}\right)^2-\frac{\Gamma}{C}} \tag{1-9}$$

With either of these values of p, either one of the Equations (1-7) gives the relation between A and B. For each value of p, one constant, say A, is arbitrary. Once A is selected, B is determined as

$$B=\frac{1}{p}A$$

Now the general solution to the homogeneous equations (1-4) is

$$x(t)=A_1e^{p_1t}+A_2e^{p_2t}$$

$$y(t)=\frac{A_1}{p_1}e^{p_1t}+\frac{A_2}{p_2}e^{p_2t} \tag{1-10}$$

with A_1 and A_2 arbitrary constants, and p_1 and p_2 given by Equation (1-9). In the special case where $\Gamma/C=(G/2C)^2$, p_1 and p_2 are equal. Then Equation (1-10) is not the solution, but the basic procedure is the same. For the present we assume $p_1 \neq p_2$ and leave the special case as an exercise (see Problem 5 at the end of the chapter).

The variation of parameters method. For the complete equation (1-3) the assumed solution is in the form of Equation (1-10) but with the free parameters A_1 and A_2 replaced by variables. Then the form is

$$v(t) = A_1(t)e^{p_1t} + A_2(t)e^{p_2t}$$
$$\lambda(t) = \frac{1}{p_1} A_1(t)e^{p_1t} + \frac{1}{p_2} A_2(t)e^{p_2t} \tag{1-11}$$

The functions $A_1(t)$ and $A_2(t)$ are determined by substituting into Equation (1-3). Thus

$$\dot{A}_1(t)e^{p_1t} + p_1A_1(t)e^{p_1t} + \dot{A}_2(t)e^{p_2t} + p_2A_2(t)e^{p_2t}$$
$$= -\frac{G}{C}[A_1(t)e^{p_1t} + A_2(t)e^{p_2t}]$$
$$-\frac{\Gamma}{C}\left[\frac{A_1(t)}{p_1}e^{p_1t} + \frac{A_2(t)}{p_2}e^{p_2t}\right] + \frac{1}{C}i(t) \tag{1-12}$$
$$\frac{1}{p_1}\dot{A}_1(t)e^{p_1t} + A_1(t)e^{p_1t} + \frac{1}{p_2}\dot{A}_2(t)e^{p_2t} + A_2(t)e^{p_2t}$$
$$= A_1(t)e^{p_1t} + A_2(t)e^{p_2t}$$

The first of these two equations can be regrouped as

$$\dot{A}_1(t)e^{p_1t} + \dot{A}_2(t)e^{p_2t} = -A_1(t)e^{p_1t}\left[p_1 + \frac{G}{C} + \frac{1}{p_1}\frac{\Gamma}{C}\right]$$
$$-A_2(t)e^{p_2t}\left[p_2 + \frac{G}{C} + \frac{1}{p_2}\frac{\Gamma}{C}\right]$$
$$+\frac{1}{C}i(t)$$

Since p_1 and p_2 are both roots of the characteristic polynomial (1-8), the contents of the two brackets are zero. With this fact and the cancellation in the second equation of (1-12), the result is

$$e^{p_1t}\dot{A}_1(t) + e^{p_2t}\dot{A}_2(t) = \frac{1}{C}i(t)$$
$$\frac{e^{p_1t}}{p_1}\dot{A}_1(t) + \frac{e^{p_2t}}{p_2}\dot{A}_2(t) = 0 \tag{1-13}$$

This is a pair of linear algebraic equations with $\dot{A}_1(t)$ and $\dot{A}_2(t)$ as unknowns. They can be solved by Cramer's rule to give

$$\dot{A}_1(t) = \frac{\begin{vmatrix} \frac{1}{C}i(t) & e^{p_2t} \\ 0 & \frac{e^{p_2t}}{p_2} \end{vmatrix}}{\begin{vmatrix} e^{p_1t} & e^{p_2t} \\ \frac{e^{p_1t}}{p_1} & \frac{e^{p_2t}}{p_2} \end{vmatrix}} = \frac{\frac{e^{p_2t}}{C\,p_2}i(t)}{e^{p_1t}e^{p_2t}\left(\frac{1}{p_2} - \frac{1}{p_1}\right)}$$

$$\dot{A}_2(t) = \frac{-\dfrac{e^{p_1 t}}{C\,p_1}\,i(t)}{e^{p_1 t}e^{p_2 t}\left(\dfrac{1}{p_2} - \dfrac{1}{p_1}\right)}$$

These forms simplify to

$$\begin{aligned} \dot{A}_1(t) &= \frac{p_1 e^{-p_1 t}}{C(p_1 - p_2)}\,i(t) \\ \dot{A}_2(t) &= \frac{p_2 e^{-p_2 t}}{C(p_2 - p_1)}\,i(t) \end{aligned} \tag{1-14}$$

By integrating Equations (1-14) the two functions $A_1(t)$ and $A_2(t)$ in (1-11) are determined to within a constant—the constant of integration. The form is more useful if the integrals are written as definite integrals. For convenience we start both integrals at t_0. This notation fits in nicely with the concept of state at t_0 defined above. Thus Equation (1-11) becomes

$$\begin{aligned} v(t) &= \left[\int_{t_0}^{t} \frac{p_1 e^{-p_1\tau}}{C(p_1 - p_2)}\,i(\tau)\,d\tau + K_1\right] e^{p_1 t} \\ &\quad + \left[\int_{t_0}^{t} \frac{p_2 e^{-p_2\tau}}{C(p_2 - p_1)}\,i(\tau)\,d\tau + K_2\right] e^{p_2 t} \\ \lambda(t) &= \left[\int_{t_0}^{t} \frac{e^{-p_1\tau}}{C(p_1 - p_2)}\,i(\tau)\,d\tau + \frac{K_1}{p_1}\right] e^{p_1 t} \\ &\quad + \left[\int_{t_0}^{t} \frac{e^{-p_2\tau}}{C(p_2 - p_1)}\,i(\tau)\,d\tau + \frac{K_2}{p_2}\right] e^{p_2 t} \end{aligned}$$

Since the integrals are with respect to τ, the functions $e^{p_i t}$ can be moved under their respective integral signs. Then the form is

$$\begin{aligned} v(t) &= \int_{t_0}^{t} \frac{p_1 e^{p_1(t-\tau)} - p_2 e^{p_2(t-\tau)}}{C(p_1 - p_2)}\,i(\tau)\,d\tau + K_1 e^{p_1 t} + K_2 e^{p_2 t} \\ \lambda(t) &= \int_{t_0}^{t} \frac{e^{p_1(t-\tau)} - e^{p_2(t-\tau)}}{C(p_1 - p_2)}\,i(\tau)\,d\tau + \frac{K_1}{p_1}\,e^{p_1 t} + \frac{K_2}{p_2}\,e^{p_2 t} \end{aligned} \tag{1-15}$$

We note that the terms involving the constants K_1 and K_2 are solutions to the homogeneous equation. Thus, the method of variation of parameters has determined $v(t)$ and $\lambda(t)$ to within a solution of the homogeneous equation. This is the best that can be done without some additional information about the solution. From the form of Equations (1-15) we see that the information is exactly what we called the state at t_0. In this particular case, if $v(t_0)$ and $\lambda(t_0)$ are known, then K_1 and K_2 are determined. If $i(t_0)$ is finite, then the integrals in Equations (1-15) are zero at $t = t_0$. Thus

$$v_1(t_0) = K_1 e^{p_1 t_0} + K_2 e^{p_2 t_0}$$

$$\lambda(t_0) = K_1 \frac{e^{p_1 t_0}}{p_1} + K_2 \frac{e^{p_2 t_0}}{p_2}$$

From these two equations K_1 and K_2 can be determined by Cramer's rule. Consequently, for this system the state at t_0 is simply two numbers.

In the next chapter the above techniques are generalized to n normal form equations with n unknowns. Before going to the generalizations, let us look at the system response (equation solution) and define a few more terms. Each of the output signals, $v(t)$ and $\lambda(t)$, contains two types of terms. One is a solution to the homogeneous equation, the other is a convolution[2] of the solution of the homogeneous equation with the input. The solution to the homogeneous equation is often called the *natural response* or the *zero-input response*. The convolution terms are called the *forced response* or *zero-state response*.

PROBLEMS

1-1 Consider the input-output relation for the circuit of Figure P1-1. Apply the definition of linearity to the relation between $v_1(t)$ and $v_2(t)$ and show that the circuit is linear.

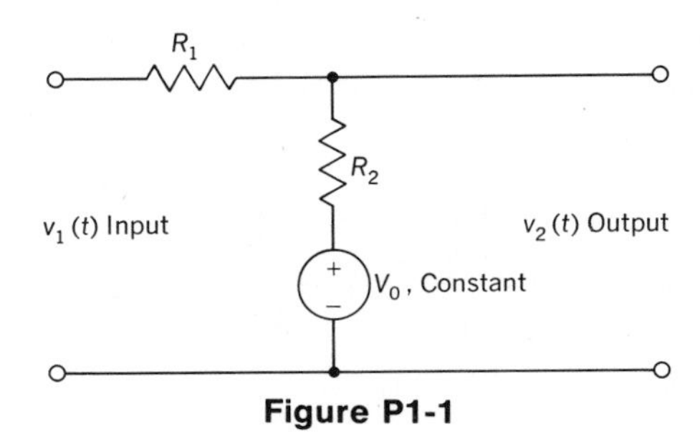

Figure P1-1

1-2 For a mass that varies with time, Newton's law is

$$f = \frac{d}{dt}(Mv)$$

where f is the force, M the mass, and v the velocity. Consider this equation as the equation of a system with velocity input and force output. Show that when M varies with time, the system is not time-invariant.

---

[2] Recall that a convolution of two functions $f(t)$ and $g(t)$ is the integral $\int_{t_0}^{t} f(t-\tau)\, g(\tau)\, d\tau$. Integrals of this form are discussed in Chapter 4.

1-3 Use the method of variation of parameters to find the current, $i(t)$, and the voltage across the capacitor, $v_0(t)$, in the first-order circuit of Figure P1-3. Both expressions should include the natural and forced response terms for any integrable input voltage $v(t)$ and initial state at t_0.

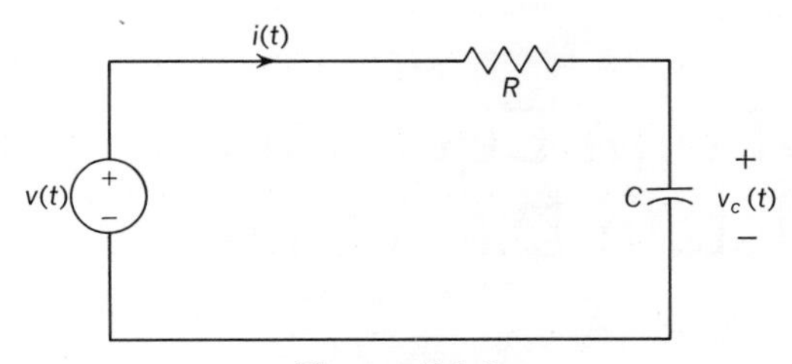

Figure P1-3

1-4 The voltage and the flux linkage for the circuit of Figure 1-1 are given by Equations (1-15). Find similar expressions for the current through the capacitor and the current through the inductor. These expressions should not require the derivative of the input current.

1-5 Find the response $v(t)$ for the circuit of Figure 1-1 when $(G/2C)^2 = \Gamma/C$. Then the solution to the homogeneous equation is $x(t) = A_1e^{pt} + A_2te^{pt}$ in Equation (1-10). Follow the subsequent steps of Section 1.3 and obtain $v(t)$ in the form

$$v(t) = \int_{t_0}^{t} [V_1e^{p(t-\tau)} + V_2(t-\tau)e^{p(t-\tau)}]\, i(\tau)\, d\tau + K_1e^{pt} + K_2te^{pt}$$

where V_1 and V_2 are constants to be determined by the method. K_1, K_2, and p are as in Equation (1-15).

CHAPTER

2

Normal Form Equations from System Diagrams

2-1 Introduction

The starting point for system analysis in this text is the *system diagram*. There are two basic diagram types that are common in the system and network literature. One is the circuit diagram, where the interconnection rules are governed by Kirchhoff's laws or D'Alembert's principle. The other is the block diagram and various related flow graphs, where the interconnection rules are summations at branch junctions. In their most elemental form the circuit diagram is an interconnection of two terminal (one-port) elements; and the block diagram is an analog computer diagram—an interconnection of integrators and amplifiers. From these elemental forms there are straightforward procedures for going to a system of first-order differential equations. For systems of more complex circuit elements or functional blocks, straightforward procedures can be derived, but their derivation requires knowledge of the forms of the solution to the differential equations. Thus, in this chapter we discuss equation formulation for the elemental diagram forms. In Chapter 4 we discuss more involved diagrams.

2-2 The Electric Circuit Diagram

The electric circuit elements that we allow in this initial formulation are the resistance, R, the capacitance, C, the inductance, L, the mutual inductance, M, the voltage controlled voltage source, μ, the current controlled voltage source, r_m, the voltage controlled current source, g_m, the current controlled current source, α, the independent voltage source, $v(t)$, and the independent current source, $i(t)$. In the present formulation all elements are linear; that is, the voltage-current relation for the resistor is $v = Ri$. The elements may be time-variant so far as the formulation is concerned. For example, R may take on different values at different times; then we would write $v(t) = R(t)\, i(t)$. In these time-variant cases we must require that the inductance and capacitance are nonzero for all time, since we divide by L and C.

With the set of allowed elements, the traditional loop or node analysis is certainly adequate for constructing a set of equilibrium equations. Thus, one might ask "Why introduce this new technique based on normal form equations?" There are several reasons. One is that the formulation of normal form equations from a circuit gives an easy way to relate the maximum number of poles in any transfer function to the number of reactive elements in the circuit. Next, the transfer function pole locations are easily computed by a classical matrix computation—the determination of the eigenvalues. There are other advantages also. Some will appear in the present development; others are beyond the scope of this text. New applications are still appearing in research papers.

Normal form equations are a set of coupled first-order equations [such as Equation (1-3)] arranged as

$$\dot{x}_i = \sum_{j=1}^{n} a_{ij}\, x_j + u_i \qquad i = 1, 2 \cdots n \tag{2-1}$$

Since differentiation of circuit variables (voltages and currents) occurs when there are inductors and capacitors, we focus our attention on these elements in formulating normal form equations from a circuit diagram.

The best method for formulating normal form equations for a circuit is based on a paper by Bashkow (Reference 1). More recent papers (see the bibliography in Reference 17) have generalized Bashkow's method somewhat, but there is still no straightforward method that handles the most general interconnection of the circuit elements listed above. Four special cases, which are discussed separately below, must be excluded from the basic procedure. They are:

1. A circuit containing a loop of capacitors.
2. A circuit containing mutual inductance or a cut set of inductors.

3. A circuit wherein capacitors and voltage sources (independent or controlled) form a loop.

4. A circuit wherein inductors and current sources (independent or controlled) form a cut set.

When the branches of an electric circuit are separated into a tree and a cotree (or tree complement or link set),[1] then Kirchhoff's current law (*KCL*) expresses tree branch currents in terms of link currents and Kirchhoff's voltage law (*KVL*) expresses link voltages in terms of tree branch voltages. When a capacitor is a tree branch, then the associated *KCL* equation is

$$C\dot{v} = \text{sum of link currents}$$

Similarly, when an inductor is a link, the *KVL* equation is

$$L\dot{i} = \text{sum of tree branch voltages}$$

Since tree branch voltages and link currents are the independent variables of the circuit, these equations are readily put in the normal form, Equation (2-1). One merely divides by the constant L or C as the case may be, and then eliminates any surplus variables as described below.

The basic procedure illustrated below by an example is as follows:

1. Select the tree called a *normal* or *proper tree* so that capacitors and voltage sources (both independent and controlled) are tree branches and inductors and current sources are links. The four restrictions above guarantee that this can be done. The resistors can be assigned to tree or cotree in any convenient manner.

2 a. For each capacitor use *KCL* to write an equation in the form

$$C\dot{v}_C = \text{sum of link currents}$$

b. For each inductor use *KVL* to write

$$L\dot{i}_L = \text{sum of tree branch voltages}$$

c. For each tree branch resistor use *KCL* to write

$$\frac{1}{R}\, v_G = \text{sum of link currents}$$

d. For each link resistor use *KVL* to write

$$Ri_R = \text{sum of tree branch voltages}$$

e. For each voltage controlled source use *KVL* to write the control voltage v_μ in terms of tree branch voltages

$$v_\mu = \text{sum of tree branch voltages}$$

---

[1] We assume the reader is familiar with some of the elementary concepts of network topology as found in most introductory circuit texts. See, for example, Reference 7.

f. For each current controlled source use KCL to write the control current i_α in terms of link currents

$$i_\alpha = \text{sum of link currents}$$

3. Consider the equations under 2e and 2f together. When a tree branch voltage in 2e is a controlled source voltage, write it in terms of its control. When a link current in 2f is a controlled source current, write it in terms of its control. Solve the resulting equations for the control variables in terms of resistive tree branch voltages, capacitor voltages, independent source voltages, resistive link currents, inductor currents, and independent source currents. The equations are independent if there is no inconsistency or indeterminateness[2] in the source specification.
4. Consider the equations under 2c and 2d together. For each controlled source current appearing on the right of 2c, write that current in terms of the various independent quantities as computed in step 3. Make the corresponding substitution for controlled source voltages in 2d. Solve the resulting equations for the tree branch resistor voltages, v_G, and the link resistor currents, i_R, in terms of capacitor voltages, inductor currents, and independent source voltages and currents. The equations are independent and can be solved because of the constraints of the network topology.
5. Substitute the results of steps 3 and 4 into the equations of 2a and 2b. The result is a set of normal form equations whose number is equal to the number of energy storage elements (inductors and capacitors) in the circuit.

The above procedure is best illustrated by an example. Consider the circuit of Figure 2-1. According to step 1 there are two possible choices of trees. One is indicated by the heavy lines in the figure; the other includes branch 7 in the tree and 4 in the cotree. The arrows on the branches indicate assumed current directions. Assumed voltage polarities

---

[2] An example of an inconsistant or indeterminate circuit is the current controlled voltage source with the control being the current through a resistor directly across the source.

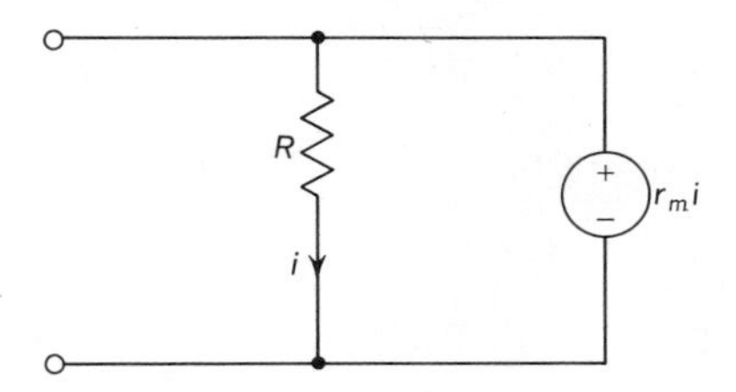

If $r_m \neq R$, the circuit is inconsistant. If $r_m = R$, any voltage and current at the terminals satisfy the element definitions and the circuit is indeterminant.

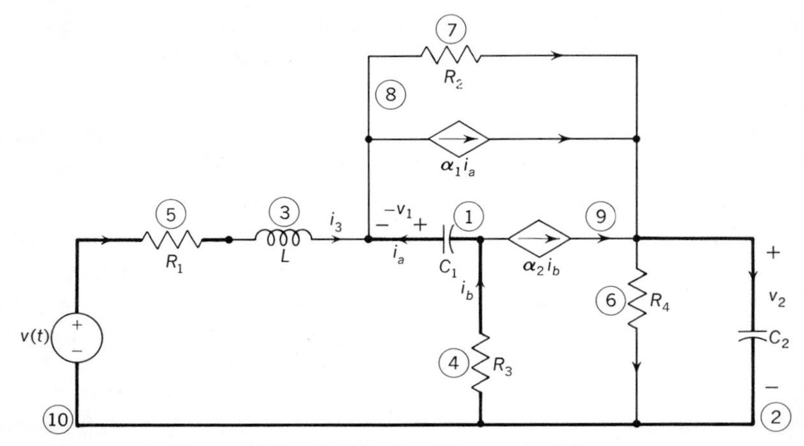

Figure 2-1 A circuit with no special cases.

are determined by the passive sign convention.[3] The numbers on branches will be used as subscripts. The equations in the order of step 2 above are

$$C_1\dot{v}_1 = -i_3 + i_7 + i_8 \tag{2-2a}$$

$$C_2\dot{v}_2 = -i_6 + i_7 + i_8 + i_9 \tag{2-2b}$$

$$L\,\dot{i}_3 = v_1 + v_4 + v_{10} - v_5 \tag{2-2c}$$

$$\frac{1}{R_3}v_4 = -i_3 + i_7 + i_8 + i_9 \tag{2-3a}$$

$$\frac{1}{R_1}v_5 = i_3 \tag{2-3b}$$

$$R_4 i_6 = v_2 \tag{2-3c}$$

$$R_2 i_7 = -v_1 - v_2 - v_4 \tag{2-3d}$$

$$i_a = -i_3 + i_7 + i_8 \tag{2-4a}$$

$$i_b = -i_3 + i_7 + i_8 + i_9 \tag{2-4b}$$

For step 3 we take Equations (2-4) and substitute for i_8 and i_9. Thus,

---

[3] Passive sign convention means voltage choosen so that when v and i have the same sign, the element absorbs power.

$$i_a = -i_3 + i_7 + \alpha_1 i_a$$

$$i_b = -i_3 + i_7 + \alpha_2 i_b + \alpha_1 i_a$$

Solving gives

$$i_a = \frac{-i_3 + i_7}{1 - \alpha_1} \tag{2-5a}$$

$$i_b = \frac{(-i_3 + i_7)}{(1 - \alpha_1)(1 - \alpha_2)} \tag{2-5b}$$

For step 4 we use Equations (2-5) in (2-3) to obtain

$$\frac{1}{R_3} v_4 = -i_3 + i_7 + \alpha_1 \left(\frac{-i_3 + i_7}{1 - \alpha_1}\right) + \alpha_2 \left(\frac{-i_3 + i_7}{(1 - \alpha_1)(1 - \alpha_2)}\right) \tag{2-6a}$$

$$\frac{1}{R_1} v_5 = i_3 \tag{2-6b}$$

$$R_4 i_6 = v_2 \tag{2-6c}$$

$$R_2 i_7 = -v_1 - v_2 - v_4 \tag{2-6d}$$

To complete step 4, Equations (2-6a) and (2-6d) must be solved simultaneously to give v_4 and i_7 in terms of v_1, v_2, and i_3. Thus

$$v_4 = -\frac{R_2 R_3}{R_3 + R_2(1 - \alpha_1)(1 - \alpha_2)} i_3 - \frac{R_3}{R_3 + R_2(1 - \alpha_1)(1 - \alpha_2)} (v_1 + v_2) \tag{2-7a}$$

$$v_5 = R_1 i_3 \tag{2-7b}$$

$$i_6 = \frac{1}{R_4} v_2 \tag{2-7c}$$

$$i_7 = \frac{R_3}{R_3 + R_2(1 - \alpha_1)(1 - \alpha_2)} i_3 - \frac{(1 - \alpha_1)(1 - \alpha_2)}{R_3 + R_2(1 - \alpha_1)(1 - \alpha_2)} (v_1 + v_2) \tag{2-7d}$$

For step 5 Equations (2-7) and (2-4) are substituted into Equation (2-2) to give the normal form equations:

$$\dot{v}_1 = -\frac{(1 - \alpha_2)}{C_1 [R_3 + R_2(1 - \alpha_1)(1 - \alpha_2)]} v_1 - \frac{(1 - \alpha_2)}{C_1 [R_3 + R_2(1 - \alpha_1)(1 - \alpha_2)]} v_2 - \frac{R_2(1 - \alpha_2)}{C_1 [R_3 + R_2(1 - \alpha_1)(1 - \alpha_2)]} i_3 \tag{2-8a}$$

$$\dot{v}_2 = -\frac{1}{C_2\,[R_3 + R_2(1-\alpha_1)(1-\alpha_2)]}v_1 - \frac{R_3 + R_4 + R_2(1-\alpha_1)(1-\alpha_2)}{C_2R_4\,[R_3 + R_2(1-\alpha_1)(1-\alpha_2)]}v_2 + \frac{1}{C_2(1-\alpha_1)(1-\alpha_2)}\left[\frac{R_3}{R_3 + R_2(1-\alpha_1)(1-\alpha_2)} - (\alpha_1 + \alpha_2 - \alpha_1\alpha_2)\right]i_3 \quad \text{(2-8b)}$$

$$\dot{i}_3 = \frac{R_2(1-\alpha_1)(1-\alpha_2)}{L[R_3 + R_2(1-\alpha_1)(1-\alpha_2)]}v_1 - \frac{R_3}{L[R_3 + R_2(1-\alpha_1)(1-\alpha_2)]}v_2 - \frac{1}{L}\left[\frac{R_2}{R_3 + R_2(1-\alpha_1)(1-\alpha_2)} + R_1\right]i_3 + \frac{v(t)}{L} \quad \text{(2-8c)}$$

Equations (2-8) are in the normal form, (2-1). The variable correspondence is $x_1 = v_1, x_2 = v_2, x_3 = i_3$. The constants a_{ij} are readily identified as the various coefficients. Finally $u_1 = 0, u_2 = 0, u_3 = v(t)/L$.

The above example illustrates a procedure for constructing normal form equations, subject to the restrictions on pages 15 and 16. When a circuit contains a loop of capacitors or a cut set of inductors, then the tree cannot be selected as in the first step of the procedure. For each capacitor loop there must be one capacitive link, and for each inductor cut set there will be an inductive tree branch. The procedure is readily corrected to take care of these special cases. Since the two situations are dual we need only consider one. Let us work out the details for a loop of capacitors.

Exceptional cases. When there is a loop of capacitors, then one link current is the current through a capacitor. This current can be written in terms of the derivative of the voltage. The voltage in turn can be written as a sum of tree branch voltages. Since the link capacitor is in a loop of capacitors, the tree branch voltages involved are all voltages across capacitors. Thus there will be no further expression of these quantities in terms of other quantities in steps 3 and 4 of the above procedure.

Let us now modify the procedure to account for the capacitor loop. In step 1, designate one capacitor in each loop of capacitors as a link. Assign the rest of the capacitors to the tree and use the same rule as before for other branches. In step 2a, the equations as stated above are for tree branch capacitors only. In steps 3 and 4 the link capacitor currents must be added to the list of variables. Step 5 must be augmented as follows: Substitute the results of 3 and 4 into the equations of 2a and 2b. Whenever a link capacitor current appears, substitute $C\,[(d/dt)$ (sum of the voltages across the other capacitors in the loop)]. Regroup the equations so that all derivatives of capacitor voltages and inductor currents are on the left and these voltages and currents plus independent source quantities are on the right. Solve these equations for the derivative terms. The solutions will be normal form equations. The network topology guarantees the independence of the equations to be solved.

To show the modified procedure consider the circuit of Figure 2-2.

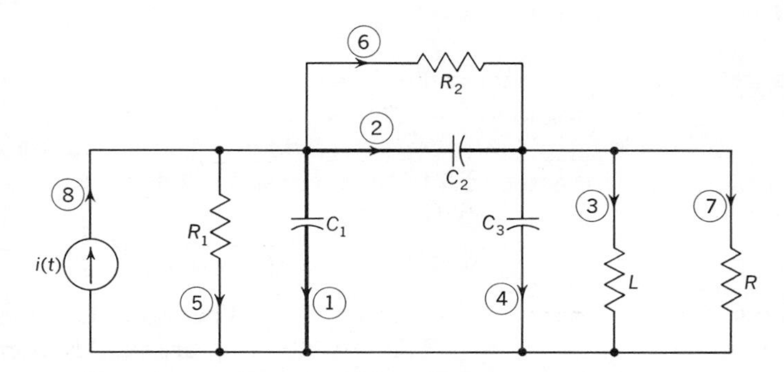

Figure 2-2 Circuit with a capacitance loop.

Since there are no controlled sources we can proceed directly to step 4 and combine with 2. Thus, the first part of step 5 is

$$C_1\dot{v}_1 = -\frac{1}{R_1}v_1 + i(t) - i_4 - i_3 - \frac{v_1 - v_2}{R_3}$$

$$C_2\dot{v}_2 = -\frac{1}{R_2}v_2 + i_4 + i_3 + \frac{v_1 - v_2}{R_3}$$

$$L\,\dot{i}_3 = v_1 - v_2$$

For i_4 we substitute $C_3(\dot{v}_1 - \dot{v}_2)$. After regrouping the result is

$$(C_1 + C_3)\,\dot{v}_1 - C_3\,\dot{v}_2 = -\left(\frac{1}{R_1} + \frac{1}{R_3}\right)v_1 + \frac{1}{R_3}v_2 - i_3 + i(t)$$

$$-C_3\dot{v}_1 + (C_2 + C_3)\,\dot{v}_2 = \frac{1}{R_3}v_1 - \left(\frac{1}{R_2} + \frac{1}{R_3}\right)v_2 + i_3$$

$$L\,\dot{i}_3 = v_1 - v_2$$

Solving for $\dot{v}_1$, $\dot{v}_2$, and $\dot{i}_3$ gives the normal form equations

$$\dot{v}_1 = -\frac{(C_2 + C_3)/R_1 + (C_2/R_3)}{C_1(C_2 + C_3) + C_2C_3}\,v_1 + \frac{C_2/R_3 - C_3/R_2}{C_1(C_2 + C_3) + C_2C_3}\,v_2$$

$$-\frac{C_2}{C_1(C_2 + C_3) + C_2C_3}\,i_3 + \frac{C_2 + C_3}{C_1(C_2 + C_3) + C_2C_3}\,i(t)$$

$$\dot{v}_2 = \frac{C_1/R_3 - C_3/R_1}{C_1(C_2 + C_3) + C_2C_3} v_1 - \frac{(C_1 + C_3)/R_2 + C_1/R_3}{C_1(C_2 + C_3) + C_2C_3} v_2 + \frac{C_1}{C_1(C_2 + C_3) + C_2C_3} i_3 + \frac{C_3}{C_1(C_2 + C_3) + C_2C_3} i(t)$$

$$\dot{i}_3 = \frac{1}{L} v_1 - \frac{1}{L} v_2$$

Since the special step required to handle a cut set of inductors is dual to that for the capacitor loop, the inductor case is left for the problems. When there is mutual inductance, a less involved modification of the procedure is required as the following example demonstrates:

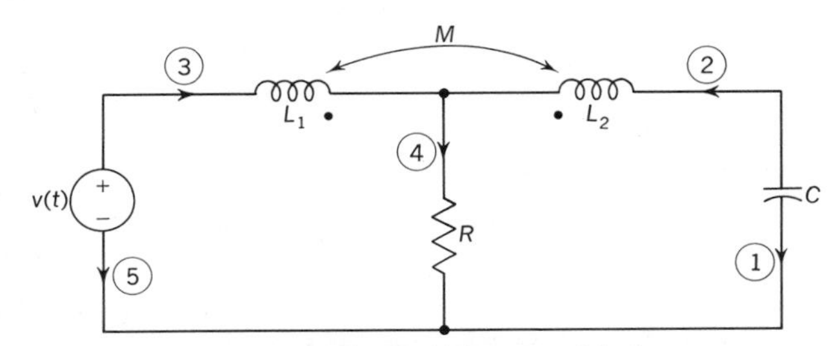

Figure 2-3 Circuit with mutual inductance.

For the circuit shown in Figure 2-3 the equations can be written immediately as

$$C\dot{v}_1 = -i_2$$

$$L_2\dot{i}_2 + M\dot{i}_3 = v_1 - (i_2 + i_3) R$$

$$L_1\dot{i}_3 + M\dot{i}_2 = -(i_2 + i_3) R + v(t)$$

The two equations involving $\dot{i}_2$ and $\dot{i}_3$ are linearly independent and can be solved simultaneously. Thus, the normal form equations are

$$\dot{v}_1 = -\frac{1}{C} i_2$$

$$\dot{i}_2 = \frac{L_1}{L_1L_2 - M^2} v_1 - \frac{R(L_1 - M)}{L_1L_2 - M^2} i_2 - \frac{R(L_1 - M)}{L_1L_2 - M^2} i_3 + \frac{M}{L_1L_2 - M^2} v(t)$$

$$\dot{i}_3 = -\frac{M}{L_1L_2 - M^2} v_1 - \frac{R(L_2 - M)}{L_1L_2 - M^2} i_2 - \frac{R(L_2 - M)}{L_1L_2 - M^2} i_3 - \frac{L_2}{L_1L_2 - M^2} v(t)$$

The last two cases that are exceptions to the basic procedure can both be illustrated by the circuit of Figure 2-4.

We recall that the unit gyrator has an impedance matrix $\begin{bmatrix} 0 & 1 \\ -1 & 0 \end{bmatrix}$ and

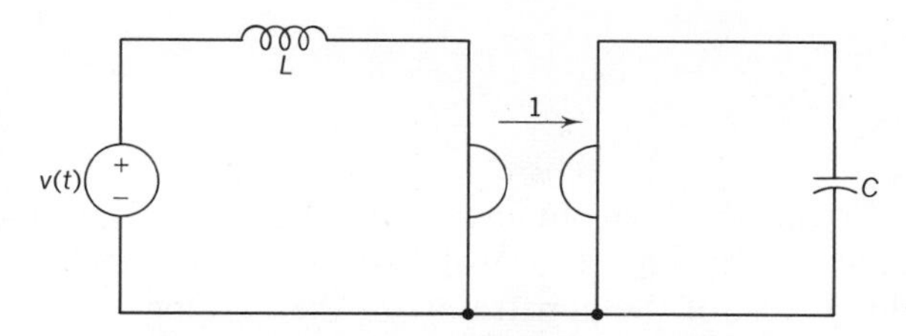

Figure 2-4 Degenerate gyrator circuit.

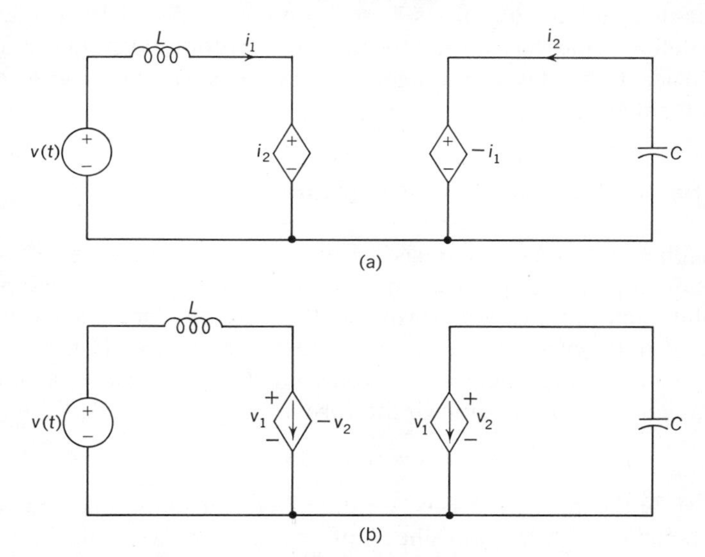

Figure 2-5 Equivalent circuits for the gyrator.

an admittance matrix that is the negative. Thus, in terms of controlled sources, Figure 2-4 can be redrawn in either of the two forms of Figure 2-5.

For Figure 2-5(a) the inductor voltage can be written

$$L\,\dot{i}_1 = v(t) - i_2$$

But

$$i_2 = C\dot{i}_1$$

Thus,

$$\dot{i}_1 = \frac{1}{L + C}\,v(t) \tag{2-9}$$

For Figure 2-5(b) the capacitor current can be written

$$C\dot{v}_2 = -v_1$$

But

$$v_1 = v(t) + L\dot{v}_2$$

Thus,

$$\dot{v}_2 = \frac{-1}{L+C} v(t) \tag{2-10}$$

Comparing Equation (2-9) obtained from Figure 2-5(a), and Equation (2-10) obtained from Figure 2-5(b), we see that they are related by $v_2 = -i_1$. This is one of the constraints of the gyrator.

It is extremely difficult to formulate a general procedure that covers all cases that violate conditions 3 and 4 on pages 15 and 16. If the circuit is well-defined and has no indeterminate condition, systematic application of Kirchhoff's laws and algebraic manipulation will lead to a set of normal form equations.

2-3 The Mechanical Circuit Diagram[4]

For mechanical circuits consisting of masses (moments of inertia), viscous damping, and springs, differentiations occur for each independent mass and each independent spring. If the variables are chosen as mass velocity v, and spring displacement y, normal form equations are readily written from a mechanical circuit diagram. For each mass, a free body diagram leads to an equation in the form

$$M\,\dot{v} = \text{a sum of } B_i v_i + K_i y_i$$

where M is the mass, the B_i's are damping constants, and the K_i's are spring constants. A careful check of spring displacements in terms of mass coordinates leads to equations of the form

$$\dot{y} = \text{a sum of } v_i$$

As an example let us consider the rotational system shown in Figure 2-6.

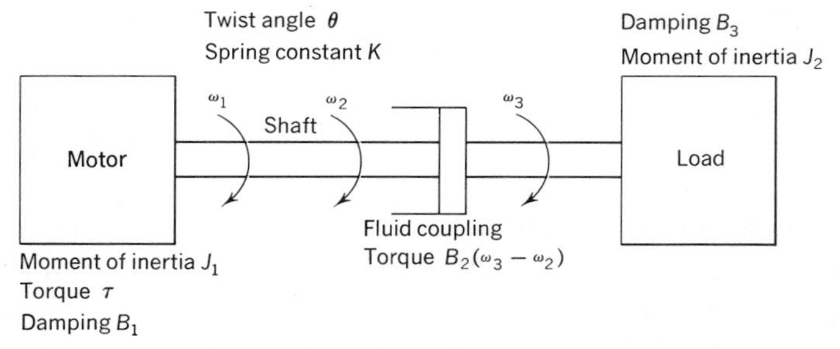

Figure 2-6 A mechanical circuit.

[4] We assume the reader is familiar with mechanical circuit diagrams as found in many introductory and intermediate circuits and systems texts. A classic reference is Reference 10.

The equations are

$$J_1\,\dot{\omega}_1 = -K\theta - B_1\,\omega_1 + \tau$$
$$J_2\,\dot{\omega}_2 = -B_3\,\omega_2 + B_2(\omega_3 - \omega_2)$$
$$\dot{\theta} = \omega_1 - \omega_3$$
$$B_2(\omega_3 - \omega_2) = K\theta$$

The last of these equations, which is an application of D'Alembert's principle where the shaft joins the fluid coupling, can be used to eliminate ω_3. Thus, the normal form equations are

$$\dot{\omega}_1 = -\frac{K}{J_1}\theta - \frac{B_1}{J_1}\,\omega_1 + \frac{1}{J_1}\,\tau$$
$$\dot{\omega}_2 = \frac{1 - B_2 - B_3}{J_2}\,\omega_2 + \frac{K}{J_2}\,\theta$$
$$\dot{\theta} = \omega_1 - \frac{1}{B_2}\,\omega_2 - \frac{K}{B_2}\,\theta$$

For more complex mechanical circuits, specific rules like those for electric circuits can be prescribed. Because of the analogy between the two systems, the rules are the same. The only problem is the transition from a sketch of mechanical devices, such as Figure 2-6, to a diagram wherein the circuit topology rules are evident. This problem is not within the scope of this text.

2-4 High-Order Equations and Analog Computer Diagrams

Often the components of a system are not specified by an interconnection of two terminal elements. Instead, each component is given by an input-output differential equation of the form

$$a_nD^nx + a_{n-1}D^{n-1}x + \cdots + a_0x = b_mD^my + \cdots + b_0y \tag{2-11}$$

where x is the output signal, the unknown; y is the input signal, a known time function; the a_i's and b_i's are coefficients of the system—they may be constants or time functions; D^i means the ith derivative with respect to time. To translate Equation (2-11) into a set of normal form equations, we first work out the homogeneous part, the left side, and then fix up the right side.

The second-order equation (1-1) was converted to two normal form, first-order equations (1-3) by defining a new variable. The same approach can convert Equation (2-11) to n normal form equations. Let

$$
\begin{aligned}
x_1 &= x \\
\dot{x}_1 &= x_2 = \dot{x} \\
\dot{x}_2 &= x_3 = \ddot{x} \\
&\vdots \\
\dot{x}_{n-1} &= x_n = D^{n-1}x
\end{aligned}
\tag{2-12}
$$

With these variables, the left side of Equation (2-11) can be written

$$a_n\dot{x}_n + a_{n-1}x_n + a_{n-2}x_{n-1} + \cdots + a_0\, x_1$$

Combining this expression with Equation (2-12) we get the relation between Equations (2-11) and (2-12), $x = x_1$, plus n normal form homogeneous equations

$$
\begin{aligned}
\dot{x}_1 &= x_2 \\
\dot{x}_2 &= x_3 \\
&\vdots \\
\dot{x}_{n-1} &= x_n \\
\dot{x}_n &= -\frac{a_0}{a_n}x_1 - \frac{a_1}{a_n}x_2 \cdots - \frac{a_{n-1}}{a_n}x_n
\end{aligned}
\tag{2-13}
$$

We can construct a procedure for accounting for the right side of Equation (2-11) by examining the effect of adding a constant times the input y to each of the equations in (2-13). Suppose we add a forcing term $K_n y$ to the last of these equations; that is,

$$
\begin{aligned}
\dot{x}_1 &= x_2 = \dot{x} \\
&\vdots \\
\dot{x}_{n-1} &= x_n = D^{n-1}x \\
x_n &= -\frac{a_0}{a_n}x_1 - \cdots - \frac{a_{n-1}}{a_n}x_n + K_n y
\end{aligned}
$$

Combining gives the nth-order equation

$$a_nD^nx + a_{n-1}D^{n-1}x + \cdots + a_0x = a_nK_ny$$

Next let us add a forcing term $K_{n-1}y$ in the next to last equation; that is,

$$
\begin{aligned}
\dot{x}_1 &= x_2 = \dot{x} \\
&\vdots \\
\dot{x}_{n-2} &= x_{n-1} = D^{n-2}x
\end{aligned}
$$

$$\dot{x}_{n-1} = x_n + K_{n-1}y = D^{n-1}x$$

$$\dot{x}_n = -\frac{a_0}{a_n}x_1 - \cdots - \frac{a_{n-1}}{a_n}x_n + K_n y = D^n x - K_{n-1}\dot{y}$$

Now the nth-order equation is

$$a_n D^n x + a_{n-1} D^{n-1}x + \cdots + a_0 x = a_n[K_n y + K_{n-1}\dot{y}] + a_{n-1}K_{n-1}y$$

From this point the pattern for the final form is clear. Thus, the equations

$$\begin{aligned}
\dot{x}_1 &= x_2 + K_1 y = \dot{x} \\
\dot{x}_2 &= x_3 + K_2 y = \ddot{x} - K_1\dot{y} \\
&\vdots \\
\dot{x}_{n-1} &= x_n + K_{n-1}y = D^{n-1}x - K_1 D^{n-2}y - \cdots K_{n-2}\dot{y} \\
\dot{x}_n &= -\frac{a_0}{a_n}x_1 - \cdots - \frac{a_{n-1}}{a_n}x_n + K_n y
\end{aligned} \tag{2-14}$$

give an nth-order equation

$$\begin{aligned}
a_n D^n x + \cdots + a_0 x = a_n &[K_1 D^{n-1}y + \cdots + K_{n-1}\dot{y} + K_n y] \\
&+ a_{n-1}[K_1 D^{n-2}y + \cdots + K_{n-2}\dot{y} + K_{n-1}y] \\
&+ \cdots + a_1 K_1 y
\end{aligned}$$

Thus, if $m \leqslant n-1$, Equations (2-11) and (2-14) are equivalent if the K_j's are given recursively by

$$\begin{aligned}
K_1 &= \frac{b_{n-1}}{a_n} \\
K_2 &= \frac{1}{a_n}\left[b_{n-2} - \frac{a_{n-1}b_{n-1}}{a_n}\right] \\
&\cdots \\
K_j &= \frac{1}{a_n}\left[b_{n-j} - a_{n-1}K_{j-1} - \cdots - a_{n-j+1}K_1\right] \\
&\cdots \\
K_n &= \frac{1}{a_n}[b_0 - a_{n-1}K_{n-1} - \cdots a_1 K_1]
\end{aligned} \tag{2-15}$$

If $m \geqslant n$, the only way to accommodate the highest-order terms is to add forcing terms involving derivatives of y in the normal form equations. Since y is known, all its derivatives are known; therefore, the resulting equations are still in normal form.

As an example of the conversion of an nth-order equation to n normal form equations consider the following third-order case:

$$\dddot{x} + 2\,\ddot{x} + 3\,\ddot{x} + 4\,x = \frac{1}{t} + \frac{2}{t^2} \tag{2-16}$$

Comparison of Equation (2-16) with (2-11) shows

$$a_0 = 4 \qquad a_1 = 3 \qquad a_2 = 2 \qquad a_3 = 1$$

If we let

$$y(t) = \frac{1}{t}$$

then

$$\dot{y}(t) - \frac{1}{t^2}$$

Now the right side of Equation (2-16) is in the form of the right side of Equation (2-11) with

$$b_0 = 1 \qquad b_1 = -2$$

From formulas (2-15)

$$K_1 = 0 \qquad K_2 = -2 \qquad K_3 = 1 + 4 = 5$$

Substituting into Equation (2-14) yields the normal form equivalent to Equation (2-16) as

$$\begin{aligned} \dot{x}_1 &= x_2 \\ \dot{x}_2 &= x_3 - \frac{2}{t} \\ \dot{x}_3 &= -4x_1 - 3x_2 - 2x_3 + \frac{5}{t} \end{aligned} \tag{2-17}$$

The connection between the variable in Equation (2-16) and those of Equations (2-17) is $x = x_1$.

To check the result we write out Equations (2-17) as in the far right side of Equations (2-14). Thus,

$$\begin{aligned} \dot{x}_1 &= x_2 = \dot{x} \\ \dot{x}_2 &= x_3 - \frac{2}{t} = \ddot{x} \\ \dot{x}_3 &= -4x_1 - 3x_2 - 2x_3 + \frac{5}{t} = \dddot{x} + \frac{d}{dt}\left(\frac{2}{t}\right) \end{aligned} \tag{2-18}$$

Substituting for x_i in the center part of Equation (2-18) gives

$$-4x - 3\dot{x} - 2\left(\ddot{x} + \frac{2}{t}\right) + \frac{5}{t} = \dddot{x} - \frac{2}{t^2}$$

Regrouping to the form of Equation (2-11) yields

$$\dddot{x} + 2\ddot{x} + 3\dot{x} + 4x = \frac{1}{t} + \frac{2}{t^2}$$

This is exactly Equation (2-16).

Equations (2-14) are not in a unique normal form equivalent to (2-11). There are infinitely many other equivalents. In Section 4-2 below some other forms are discussed. No attempt is made to catalog all useful forms; procedures similar to the one above are readily derived for any desired normal form equivalent to Equation (2-11).

Modeling on an analog computer. Equations such as (2-11) are often modeled on an analog computer in order to obtain a graphical solution, or to be used as part of a more complex simulated system. The basic elements of the computer, so far as this equation is concerned, are integrators, amplifiers, and summers. The symbols that we use for these elements along with their input-output definitions are shown in Figure 2-7. If, in addition to the above elements, we also had a differentiator, it would be trivial to model Equation (2-11) directly. The normal form equations (2-14) are just as easily diagramed using only the three elements of Figure 2-7.

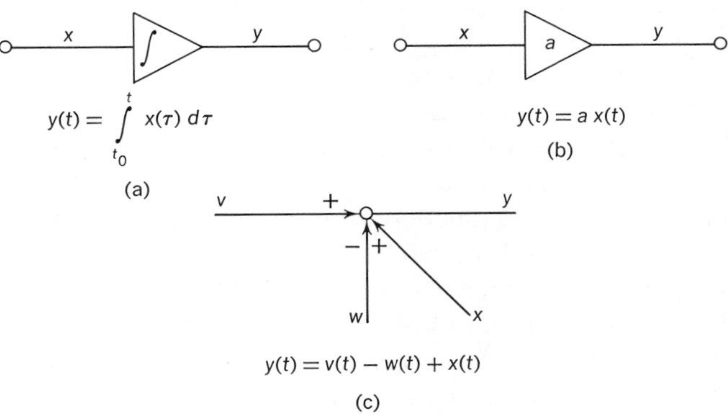

Figure 2-7 Analog diagram elements.

A procedure for modeling is derived by starting with one integrator for each of the variables x_i in (2-14). The output of the ith integrator is x_i; the input is then $\dot{x}_i$. For $i \neq n$, the ith of Equations (2-14) says that $\dot{x}_i$, the input to the ith integrator, must be (x_{i+1}), the output of the $(i + 1)$th

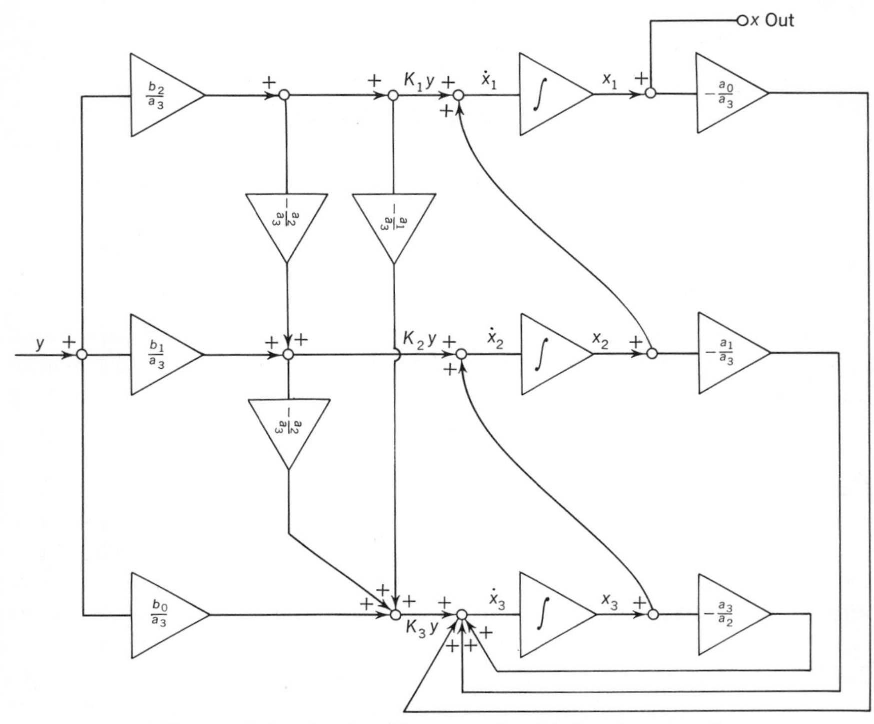

Figure 2-8 Analog diagram of a third-order equation.

integrator, plus K_i times the input. Finally, the input to the nth integrator, $\dot{x}_n$, can be formed as the appropriately weighted sum of the outputs of all the integrators plus K_n times the input. The result is shown in Figure 2-8 for $n = 3$.

The inverse problem of going from an analog computer diagram to a system of normal form equations is even more straightforward. One merely defines one variable for each integrator output. The inputs are then the derivatives of the respective variables. Then an equation is written for each summing junction. If there are junctions whose outputs lead through amplifiers (but not integrators) to other junctions, there will be additional algebraic equations. New additional variables may have to be defined for ease in constructing such equations. If the system diagram is well-defined, for each new variable there will be an independent algebraic equation that can be used to eliminate the new variable.

The final point that needs clarification for our analog computer diagrams is that of initial conditions on the integrators. The state at t_0 determines how the integrators should be set up. This matter will be left until Section 4-3 below after a general discussion of initial conditions for normal form equations.

PROBLEMS

2-1 Set up normal form equations for the circuit of Figure P2-1.

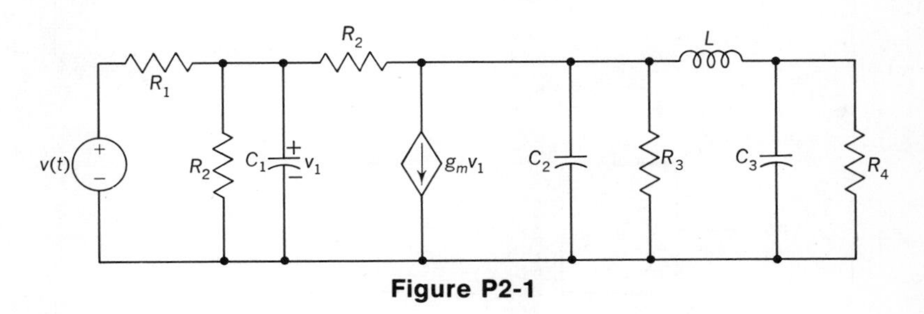

Figure P2-1

2-2 Set up normal form equations for the circuit of Figure P2-2.

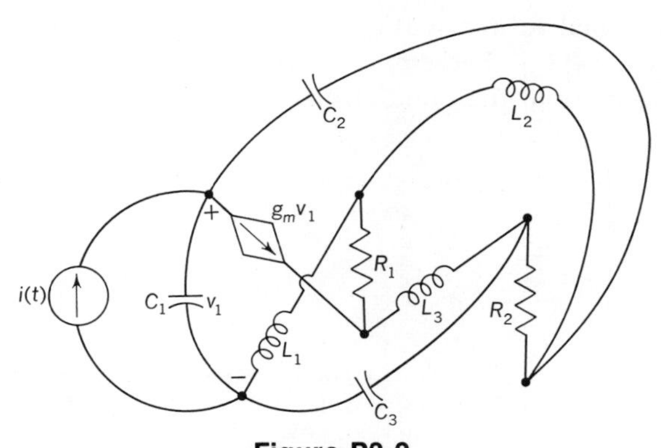

Figure P2-2

2-3 Set up normal form equations for the circuit of Figure P2-3.

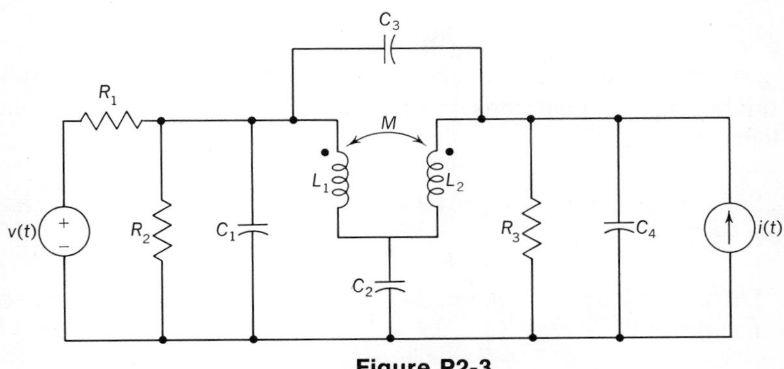

Figure P2-3

2-4 Set up normal form equations for the one-dimensional mechanical circuit of Figure P2-4.

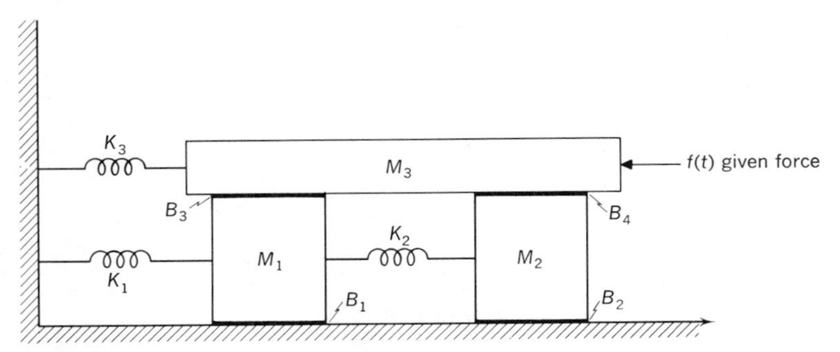

Figure P2-4 The B's represent friction along the surfaces (viscous friction, force proportional to velocity). Motion in X direction only.

2-5 Set up normal form equations for the two-dimensional mechanical system of Figure P2-5. The equations will be nonlinear. Linearize them for small motion about the equilibrium point.

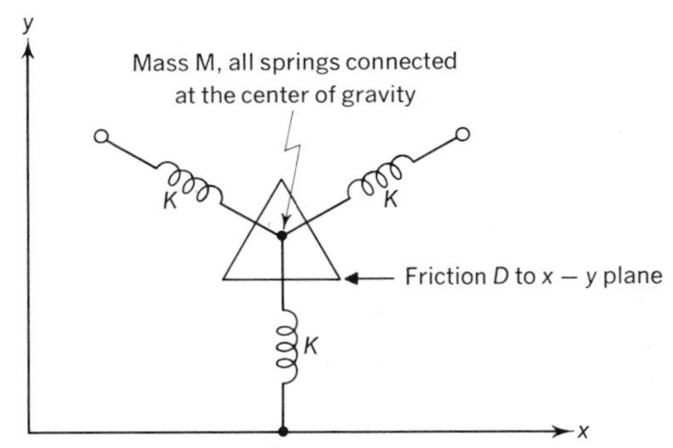

Figure P2-5 Motion in $x - y$ plane. Springs are such that the angle between them is 120 degrees at equilibrium.

2-6 Set up normal form equations for the electromechanical circuit of Figure P2-6. Since the equations for capacitance and electrical force are nonlinear, the equations are nonlinear. Linearize them about an equilibrium point x_0.

2-7 Set up normal form equations for the analog computer diagram of Figure P2-7.

2-8 Draw an analog computer diagram for the normal form equations with $\underset{\sim}{A}$ matrix $\begin{bmatrix} 0 & -1 & -2 \\ 1 & 3 & 7 \\ 4 & -2 & -1 \end{bmatrix}$

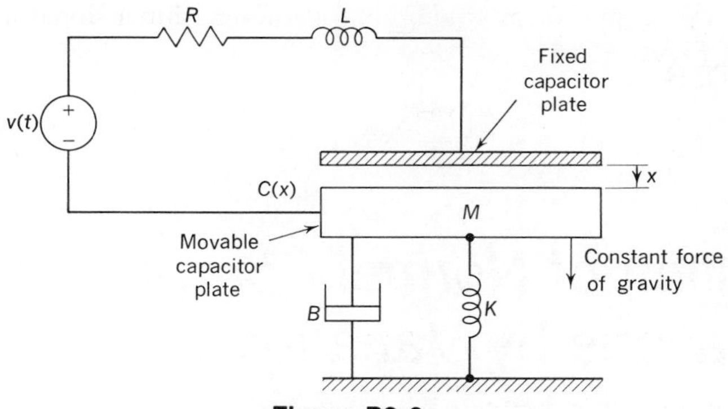

Figure P2-6

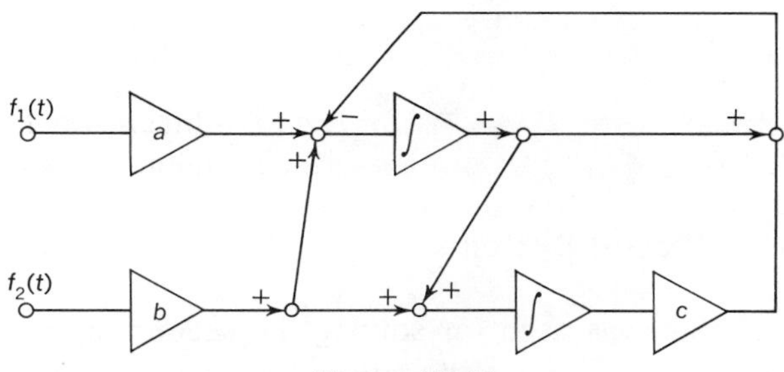

Figure P2-7

2-9 Define three state variables and set up normal form equations from the equation

$$\dddot{x} + 3\ddot{x} + 2\dot{x} + x = 3 \cos 10t + 40 \sin 10t$$

Make the input cos $10t$ times a suitable vector to give the vector forcing function $\mathbf{U}(t)$.

2-10 Define five state variables and set up normal form equations for the equations

$$\dddot{x} + 3\ddot{x} + 2\dot{x} + x + 3\ddot{y} + 2\dot{y} + y = \cos 10t$$

$$\ddot{y} + 3\dot{y} + x = \sin 10t$$

CHAPTER

3

Solution of Normal Form Equations by Variation of Parameters

3-1 Fixed Coefficient Systems

The procedure that was used for solving the second-order system (1-3) in Section 1-3 can also be used to solve the general normal form Equations (2-1) when the coefficients a_{ij} are all constants. As pointed out below, much of the method still applies when the coefficients are functions of time as well. The method is most compactly described in matrix notation. Thus, Equations (2-1) can be written as

$$\dot{\mathbf{x}} = \underset{\sim}{A}\,\mathbf{x} + \mathbf{u} \tag{3-1}$$

where $\mathbf{x}$ is an $n \times 1$ matrix (n-vector) of time functions; $\dot{\mathbf{x}}$ is an n-vector derived from $\mathbf{x}$ by differentiating each component; $\underset{\sim}{A}$ is an $n \times n$ matrix whose elements are the a_{ij} in Equation (2-1) assumed constant in this section; $\mathbf{u}$ is an n-vector of known time functions.[1]

The homogeneous equations. The first step in the solution is the solution of the homogeneous equation

$$\dot{\mathbf{w}} = \underset{\sim}{A}\,\mathbf{w} \tag{3-2}$$

where $\underset{\sim}{A}$ is the same as in Equation (3-1).

A solution to Equation (3-2) can be obtained by assuming each component of $\mathbf{w}$, w_i, has the form $W_i e^{pt}$, and then substituting into the

[1] Henceforth bold face letters will represent vectors. Letters with a tilde underneath will represent matrices.

equation. The W_i's are constants as is p. In vector notation we define a vector of constants $\mathbf{W}$ and assume

$$\mathbf{w} = e^{pt}\,\mathbf{W} \tag{3-3}$$

Substituting into Equation (3-2) gives

$$pe^{pt}\,\mathbf{W} = e^{pt}\,\underset{\sim}{A}\,\mathbf{W}$$

Cancelling the e^{pt}'s and regrouping gives the set of homogeneous algebraic equations in the W_i

$$[\underset{\sim}{A} - p\,\underset{\sim}{I}]\,\mathbf{W} = 0 \tag{3-4}$$

where $\underset{\sim}{I}$ is the $n \times n$ identity matrix.

This equation, like (1-7) above, has nontrivial solutions only if the determinant of the coefficients is zero. The determinant

$$|\underset{\sim}{A} - p\,\underset{\sim}{I}| = \begin{vmatrix} a_{11} - p & a_{12} & & a_{1n} \\ a_{21} & a_{22} - p & \cdots & a_{2n} \\ \vdots & & & \\ a_{n1} & a_{n2} & & a_{nn} - p \end{vmatrix} \tag{3-5}$$

is an nth degree polynomial in p called the *characteristic polynomial.* Its n roots are the allowable values of p for the solution (3-3) for the homogeneous equation. These values are called the *natural frequencies of the system*, or the *eigenvalues* of the $\underset{\sim}{A}$ matrix. We let $p_1, p_2, \cdots, p_n$ be these natural frequencies.

The homogeneous algebraic equations, (3-4), like (1-7), determine ratios between the components of $\mathbf{W}$. For each p_i, if one component of $\mathbf{W}$ is set arbitrarily, all other components are determined, provided p_i is not a multiple root of the characteristic polynomial. For the present development let us assume that all the roots are distinct. In the special cases when they are not, more space is needed for adequate description than they are worth in an introductory treatment. The procedures are not difficult and the student can work them out for himself. (See Problems 2-1, 3-1, and 3-2.) Furthermore, the shorter computational methods discussed below will handle the simultaneous root case directly.

The general solution to the homogeneous equation accounting for all natural frequencies is

$$\mathbf{w} = e^{p_1 t}\,\mathbf{W}_1 + e^{p_2 t}\,\mathbf{W}_2 + \cdots + e^{p_n t}\,\mathbf{W}_n \tag{3-6}$$

where the p_i's are the eigenvalues of the $\underset{\sim}{A}$ matrix of (3-1) here assumed to be distinct; the $\mathbf{W}_i$ are n-vectors whose components are determined to within an arbitrary constant multiple by the algebraic equations (3-4).

Variation of parameters. The variation of parameters procedure for the nth-order system (3-1) is exactly analogous to that for the second-order system (1-3). The solution to the complete equation (3-1) is assumed to be in the form of the general solution to the homogeneous equation (3-6), but with the arbitrary constants in Equation (3-6) replaced by functions of time to be determined by the method. Thus we assume

$$\mathbf{x} = \beta_1(t)\, e^{p_1 t}\, \mathbf{W}_1 + \beta_2(t)\, e^{p_2 t}\, \mathbf{W}_2 + \cdots + \beta_n(t)\, e^{p_n t}\, \mathbf{W}_n \tag{3-7}$$

The summation in Equation (3-7) can be written compactly in matrix notation

$$\mathbf{x} = \underset{\sim}{\phi}(t)\, \boldsymbol{\beta}(t) \tag{3-8}$$

where

$$\underset{\sim}{\phi}(t) = \begin{bmatrix} W_{11}e^{p_1 t} & W_{21}e^{p_2 t} & W_{n1}e^{p_n t} \\ W_{12}e^{p_1 t} & W_{22}e^{p_2 t} & W_{n2}e^{p_n t} \\ \vdots & \vdots & \vdots \\ W_{1n}e^{p_1 t} & W_{2n}e^{p_2 t} & W_{nn}e^{p_n t} \end{bmatrix} \tag{3-9}$$

W_{ij} is the jth component of the vector $\mathbf{W}_i$ in (3-6); $\boldsymbol{\beta}(t)$ is a vector whose components are the $\beta_i(t)$ of Equation (3-7).

The matrix $\underset{\sim}{\phi}$ as written out above satisfies the matrix differential equation

$$\dot{\underset{\sim}{\phi}} = \underset{\sim}{A}\, \underset{\sim}{\phi} \tag{3-10}$$

This statement is readily verified by writing out the matrices of (3-10) and noting that each column is a statement of the homogeneous differential equation (3-2) with a p_i that is a root of the characteristic polynomial (3-5), and $\mathbf{W}_{ij}$ that satisfies (3-4) with that p_i. In fact, the variation of parameters procedure below works with any matrix $\underset{\sim}{\phi}$ that satisfies (3-10) and whose columns are linearly independent solutions to the homogeneous equation (3-2). Any such matrix is called a *fundamental matrix* for the homogeneous system (3-2). We use the symbol $\underset{\sim}{\phi}(t)$ for any fundamental matrix. One example is Equation (3-9). We will discuss others below. Since the columns of $\underset{\sim}{\phi}$ are linearly independent, its determinant is nonzero and thus the inverse matrix $\underset{\sim}{\phi}^{-1}$ exists.

Substituting Equation (3-8) into (3-1) gives

$$\dot{\underset{\sim}{\phi}}\, \boldsymbol{\beta} + \underset{\sim}{\phi}\, \dot{\boldsymbol{\beta}} = \underset{\sim}{A}\, \underset{\sim}{\phi}\, \boldsymbol{\beta} + \mathbf{u} \tag{3-11}$$

By Equation (3-10) much of Equation (3-11) can be eliminated; thus,

$$\underset{\sim}{\phi}\, \dot{\boldsymbol{\beta}} = \mathbf{u} \tag{3-12}$$

This is a set of linear algebraic equations with the components of $\dot{\boldsymbol{\beta}}$ as unknowns. Since $\underset{\sim}{\phi}$ is a fundamental matrix, the determinant of the coefficients is nonzero and Equation (3-12) can be solved for these unknowns. Symbolically,

$$\dot{\boldsymbol{\beta}} = \underset{\sim}{\phi}^{-1} \mathbf{u} \tag{3-13}$$

The vector $\boldsymbol{\beta}$ can be determined by integrating each of the n equations represented by (3-13). Thus, the variation of parameters determines each component of $\boldsymbol{\beta}(t)$ to within a constant of integration. Symbolically we may write

$$\boldsymbol{\beta}(t) = \int^{t} \underset{\sim}{\phi}^{-1}(\tau)\ \mathbf{u}(\tau)\ d\tau \tag{3-14}$$

where the lower limit is left off of the integral to show that the constant of integration is not yet determined. Substituting Equation (3-14) into Equation (3-8) gives the solution to Equation (3-1) as

$$\mathbf{x}(t) = \underset{\sim}{\phi}(t) \int^{t} \underset{\sim}{\phi}^{-1}(\tau)\ \mathbf{u}(\tau)\ d\tau \tag{3-15}$$

Since τ is the variable of integration, the matrix $\underset{\sim}{\phi}(t)$ can be written inside the integral as well as outside. Often it is more convenient to multiply $\underset{\sim}{\phi}(t)$ and $\underset{\sim}{\phi}^{-1}(\tau)$ before integrating. In this case, Equation (3-15) can be rewritten

$$\mathbf{x}(t) = \int^{t} \underset{\sim}{\phi}(t)\ \underset{\sim}{\phi}^{-1}(\tau)\ \mathbf{u}(\tau)\ d\tau \tag{3-16}$$

For the nth-order system characterized by (3-1), the response (3-15) has n constants of integration. Virtually any n-independent conditions of the system can be used to determine these constants. One possibility is the specification of one component of $\mathbf{x}$ at n different times. Another is the specification of one component and $(n - 1)$ of its derivatives at a particular time. The most convenient specification for many practical problems and mathematical problems is the specification of all components of $\mathbf{x}$ at a particular time t_0. Such a specification is closely related to the definition of the state at t_0 as defined above.

The state transition matrix. For each particular system specification for determining the constants, there is a particular form of the fundamental matrix that is most convenient for translating system conditions to constants of integration in formula (3-15) for the response. We shall discuss only the form that is convenient when each component of $\mathbf{x}$ is specified at t_0; that is, $\mathbf{x}(t_0)$ is known. If we change the indefinite integrals in (3-15) to definite integrals all starting at t_0, then the response has the form

$$\mathbf{x}(t) = \underset{\sim}{\phi}(t) \int_{t_0}^{t} \underset{\sim}{\phi}^{-1}(\tau)\ \mathbf{u}(\tau)\ d\tau + \underset{\sim}{\phi}(t)\ \mathbf{K} \tag{3-17}$$

where **K** is a vector of constants. At t_0, Equation (3-17) becomes

$$\mathbf{x}(t_0) = \underset{\sim}{\phi}(t_0)\ \mathbf{K} \tag{3-18}$$

If we can find a fundamental matrix, such that $\underset{\sim}{\phi}(t_0)$ is the $n \times n$ identity matrix, then in (3-18), **K** is equal to $\mathbf{x}(t_0)$. Such a matrix can always be found as shown below. This fundamental matrix is called the *state transition matrix*. We use the notation $\underset{\sim}{\phi}(t, t_0)$ for this special fundamental matrix.

Any solution of the homogeneous equation (3-2) can be written as a linear combination of the solutions represented by the columns of the $\underset{\sim}{\phi}(t)$ defined by Equation (3-9). Thus, for any $\underset{\sim}{\phi}(t)$, the elements ϕ_{ij} can all be written in the form

$$\phi_{ij}(t) = \alpha_{ij1}\ e^{p_1 t} + \alpha_{ij2}\ e^{p_2 t} + \cdots + \alpha_{ijn}\ e^{p_n t} \tag{3-19}$$

where the α_{ijk} are constants and the assumption that the characteristic polynomial has distinct roots still applies.

With each ϕ_{ij} written as in (3-19), the $\underset{\sim}{\phi}$ matrix has n^3 constants. Since each column must satisfy the homogeneous equation (3-2), only n^2 of the constants are free to assume a particular form. If we wish to obtain a $\underset{\sim}{\phi}$ matrix whose entries are all constrained to have a particular value at t_0, then we have exactly the n^2 constraints needed to determine the n^2 free constants. Thus, by solving n^2 algebraic equations, we can find the state transition matrix. There are easier ways to find this matrix; some of which are discussed below.

In terms of the state transition matrix, the system variables for all time can be written

$$\mathbf{x}(t) = \int_{t_0}^{t} \underset{\sim}{\phi}(t, t_0)\ \underset{\sim}{\phi}^{-1}(\tau, t_0)\ \mathbf{u}(\tau)\ d\tau + \underset{\sim}{\phi}(t, t_0)\mathbf{x}(t_0) \tag{3-20}$$

From this expression we see that $\mathbf{x}(t_0)$ is the state at t_0. If the input signals are known between t_0 and t_1, then $\mathbf{u}(t)$ is readily computed in that interval. With $\mathbf{u}(t)$ known, the integral can be calculated. The only additional information needed is that required for the second term in Equation (3-20). The n numbers in $\mathbf{x}(t_0)$ provide that information. The state transition matrix $\underset{\sim}{\phi}(t, t_0)$ determines the change of the state from t_0 to another time—hence, the name "state transition matrix." Since **x** at any time determines the state at that time, **x** is called a *state vector*. The components of **x** vary with time and are called *state variables*. The n-dimensional space that the n-vector **x** determines is then called the *state space*. In this text the above three terms will be used descriptively.

They are not applied as mathematical definitions. From Equation (3-20) the motivation for calling the integral term the zero-state response and the second term the zero-input response is clear.

3-2 Some Remarks on Time-Variant Systems

In the preceding section the $\underset{\sim}{A}$ matrix in Equation (3-1) was assumed to be a matrix of constants. As we will show in Chapter 4, the response satisfies the definition of time-invariance given in Section (1-2). When the elements of the $\underset{\sim}{A}$ matrix are functions of time, the system is still linear, but it is no longer time-invariant—it is *time-variant*. The equation formulation methods of Chapter 2 apply when the various diagram parameters are functions of time. This case leads to an $\underset{\sim}{A}$ matrix of time functions.

As an example let us consider a third-order equation with variable coefficients similar to the constant coefficient equation (2-16). One possibility is

$$\dddot{x} + (2 \cos 10t)\,\ddot{x} + \frac{1}{t}\,\dot{x} + x = \frac{1}{t} + \frac{2}{t^2} \tag{3-21}$$

The steps that led from Equation (2-16) to (2-17) still apply. The result is

$$\dot{\mathbf{x}} = \underset{\sim}{A}(t)\,\mathbf{x} + \mathbf{u} \tag{3-22}$$

where

$$x_1 = x \text{ from (3-21)}$$

$$x_2 = \dot{x}$$

$$x_3 = \ddot{x} + \frac{2}{t}$$

$$\underset{\sim}{A}(t) = \begin{bmatrix} 0 & 1 & 0 \\ 0 & 0 & 1 \\ -1 & -\dfrac{1}{t} & -2\cos 10t \end{bmatrix}$$

$$\mathbf{u} = \begin{bmatrix} 0 \\ -\dfrac{2}{t} \\ \dfrac{1}{t} + (2\cos 10t)\,\dfrac{2}{t} \end{bmatrix}$$

The method of variation of parameters applies to all linear equations; time-variant as well as time-invariant. Nevertheless, there is no general method for finding explicit analytical solutions to the general time-variant systems problems because there is no general method for solving the homogeneous equations in closed form in terms of well-known functions. For certain special second-order equations, such as Bessel's equation and Mathieu's equation, special functions that are solutions to the homogeneous equation have been tabulated. Thus, if the homogeneous equation for a system is Bessel's equation, the response can be computed in terms of Bessel functions.

If, for an nth-order system, one had n linearly independent solutions to the homogeneous equation, a fundamental matrix could be constructed. In the time-variant case the entries are not simple exponential functions. As in the time-invariant case, there is a unique state transition matrix $\underset{\sim}{\phi}(t, t_0)$ such that $\underset{\sim}{\phi}(t_0, t_0)$ is the identity matrix. Thus, formula (3-20) applies equally well to the response of a time-variant system.

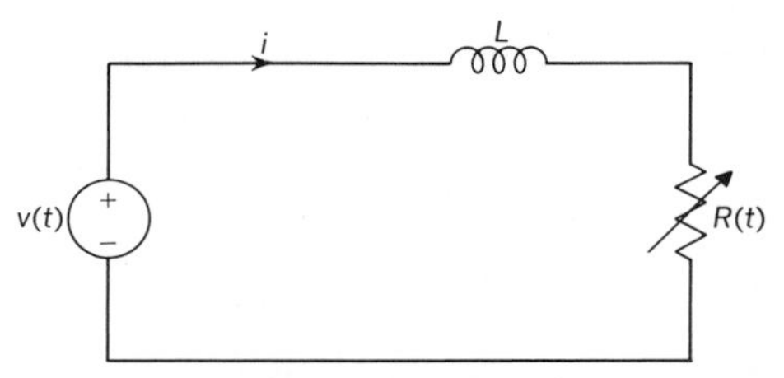

Figure 3-1 A first-order, time-variant system.

Let us illustrate the procedure with the only general case of time-variant systems that can be solved explicitly—the first-order systems. To be specific, consider the circuit of Figure 3-1. For this circuit Kirchhoff's law gives the first-order differential equation in normal form

$$\dot{i}(t) = -\frac{R(t)}{L} i(t) + \frac{v(t)}{L} \tag{3-23}$$

The homogeneous part of Equation (3-23) is

$$\dot{x}(t) = -\frac{R(t)}{L} x(t) \tag{3-24}$$

Separating variables and integrating gives

$$\ln x(t) = -\int^{t} \frac{R(t)}{L}\, d\tau$$

or

$$x(t) = \exp\left(-\int^{t} \frac{R(\tau)}{L}\, d\tau\right) \tag{3-25}$$

For this first-order system, the right side of Equation (3-25) is the general form for a fundamental matrix. The constant of integration in

(3-25) is readily chosen so that $x(t_0) = 1$. Thus, in this case, the state transition matrix is

$$\phi(t, t_0) = \exp\left(-\int_{t_0}^{t} \frac{R(\tau)}{L}\, d\tau\right) \tag{3-26}$$

For the variation of parameter step we proceed as with Equation (3-7) above and assume the solution $i(t)$ to Equation (3-23) in the form

$$i(t) = \beta(t) \exp\left(-\int_{t_0}^{t} \frac{R(\tau)}{L}\, d\tau\right) \tag{3-27}$$

Here the $\beta(t)$ is the "variable parameter" to be determined and the exponential is a solution to the homogeneous equation. Substituting Equation (3-27) into (3-23) gives

$$\dot{\beta}(t) \exp\left(-\int_{t_0}^{t} \frac{R(\tau)}{L}\, d\tau\right) + \beta(t)\left(-\frac{R(t)}{L}\right) \exp\left(-\int_{t_0}^{t} \frac{R(\tau)}{L}\, d\tau\right)$$
$$= -\frac{R(t)}{L}\beta(t) \exp\left(-\int_{t_0}^{t} \frac{R(\tau)}{L}\, d\tau\right) + \frac{v(t)}{L} \tag{3-28}$$

Solving for $\dot{\beta}(t)$ gives

$$\dot{\beta}(t) = \frac{v(t)}{L} \exp\left(\int_{t_0}^{t} \frac{R(\tau)}{L}\, d\tau\right) \tag{3-29}$$

Integrating Equation (3-29) gives

$$\beta(t) = \int_{t_0}^{t} \frac{v(\lambda)}{L}\left[\exp\left(\int_{t_0}^{\lambda} \frac{R(\tau)}{L}\, d\tau\right)\right] d\lambda + K \tag{3-30}$$

where K is a constant to be determined from initial conditions.

Substituting Equation (3-30) into (3-27) gives the solution $i(t)$ to Equation (3-23). It is

$$i(t) = \int_{t_0}^{t} \frac{v(\lambda)}{L}\left[\exp\left(\int_{t_0}^{\lambda} \frac{R(\tau)}{L}\, d\tau\right)\right]\left[\exp\left(-\int_{t_0}^{t} \frac{R(\tau)}{L}\, d\tau\right)\right] d\lambda$$
$$+ K \exp\left(-\int_{t_0}^{t} \frac{R(\tau)}{L}\, d\tau\right) \tag{3-31}$$

Since at $t = t_0$ all integrals in Equation (3-31) are zero, then $K = i(t_0)$. Rewriting Equation (3-31) in terms of $i(t_0)$ and the state transition matrix (3-26) gives

$$i(t) = \int_{t_0}^{t} \phi(t, t_0)\, \phi^{-1}(\tau, t_0)\, \frac{v(\tau)}{L}\, d\tau + i(t_0)\, \phi(t, t_0) \tag{3-32}$$

This last equation is exactly the same as Equation (3-20) for the one-dimensional case. Again the response is the sum of a zero-state term and a zero-input term.

3-3 Some Simpler Computational Techniques

The method of variation of parameters is a compact way of demonstrating the form of the response of a linear system without the need for any steps that cannot be justified by elementary mathematics. As a computational technique for computing input-output relations for linear, time-invariant systems, the method is long and tedious. In this section we establish the relation between the general response (3-20) and the input-output computational techniques used in most undergraduate circuits and systems texts.

The normal form equations (3-1) do not display the system inputs and outputs specifically. The vector **u** is linearly related to the inputs, but its components are not necessarily the source quantities exactly. In general, the n-dimensional vector **u** is related to the m sources characterized by an m-vector **v**, by an $n \times m$ matrix B. Then in terms of the source vector **v**, Equation (3-1) becomes

$$\dot{\mathbf{x}} = \underset{\sim}{A}\,\mathbf{x} + \underset{\sim}{B}\,\mathbf{v} \tag{3-33}$$

Sometimes the physical sources will be specified in a way that requires differentiation of a source quantity in the normal form equations. We shall not carry such cases specifically because the notation gets messy. The principle is the same.

The system will in general have p outputs with p not necessarily equal to m or n. For a linear system the outputs will be a linear combination of the state variables, the inputs, and derivatives, and integrals of the state variables and the inputs. For reasons of stability as defined in Chapter 9, derivatives of the input beyond $\dot{\mathbf{v}}$ are not allowed. Integrals of the input can be accounted for by modifying the definitions of the state variables. Thus, the output **y** may be written

$$\mathbf{y} = \underset{\sim}{C}\,\mathbf{x} + \underset{\sim}{D}\,\mathbf{v} + \underset{\sim}{E}\,\dot{\mathbf{v}} \tag{3-34}$$

where $\underset{\sim}{C}$ is $p \times n$ and $\underset{\sim}{D}$ and $\underset{\sim}{E}$ are $p \times m$.

The complete system with input and output is described by combining Equation (3-33) and Equation (3-34) to give

$$\begin{aligned} \dot{\mathbf{x}} &= \underset{\sim}{A}\,\mathbf{x} + \underset{\sim}{B}\,\mathbf{v} \\ \mathbf{y} &= \underset{\sim}{C}\,\mathbf{x} + \underset{\sim}{D}\,\mathbf{v} + \underset{\sim}{E}\,\dot{\mathbf{v}} \end{aligned} \tag{3-35}$$

Frequency analysis. When our interest is only in input-output relationships for a system, there is a much easier way to get the desired result than variation of parameters. The simplified computational method is often introduced via Laplace transforms. However, when transforms are used, one must be careful about regions of convergence if he wants a general analysis that is not tied to an initial starting time, such as $t = 0$. The same results can be obtained without skipping over as many mathematical points by using the functional form (3-17) for the state variables and examining a very special forcing function, the *sine wave*.

When the $\underset{\sim}{\phi}$ matrix used in Equation (3-17) has the form given by Equation (3-9), it is easy to show the functional form of the elements of $\underset{\sim}{\phi}(t)\ \underset{\sim}{\phi}^{-1}(\tau)$. The determinant of $\underset{\sim}{\phi}$ is given by a constant, the determinant of the W_{ij}'s, times the product of all the $e^{p_i t}$. Each cofactor of the element ϕ_{ij} is a cofactor of the matrix of the W_{ij}'s times the product of all the $e^{p_i t}$ except $e^{p_j t}$. Thus, the i, j element of $\phi^{-1}(t)$ is

$$\phi_{ij}^{-1}(t) = \frac{|\underset{\sim}{W}_{ji}|}{|\underset{\sim}{W}|} e^{-p_i t} \tag{3-36}$$

where

$|\underset{\sim}{W}_{ji}|$ is the cofactor (sign included) of the ji element in the $\underset{\sim}{W}$ matrix
$|\underset{\sim}{W}|$ is the determinant of the $\underset{\sim}{W}$ matrix

With $\phi_{ij}^{-1}(t)$ having the form (3-36) and $\phi_{ij}(t)$ from Equation (3-9) having the form $W_{ij}e^{p_i t}$, the ij element of $\phi(t)\ \phi^{-1}(\tau)$ has the form

$$\sum_{k=1}^{n} K_{ijk}\, e^{p_k(t-\tau)} \tag{3-37}$$

Now the forced (zero-state) response term for each state variable takes the form

$$x_i(t) = \sum_{j=1}^{n} \int_{t_0}^{t} \sum_{k=1}^{n} K_{ijk}\, e^{p_k(t-\tau)}\, u_j(\tau)\, d\tau \tag{3-38}$$

where the u_j are the elements of $\mathbf{u}$ in Equation (3-1). When there are multiple roots to the characteristic polynomial the derivation to a functional form equivalent to (3-38) is straightforward, but messy. The result is the possible inclusion of terms of the form

$$K_{ijk}(t-\tau)^l\, e^{p_m(t-\tau)}\, u_j(\tau) \tag{3-39}$$

where $l + m = k$ and l takes on values $0, 1, 2, \cdots (h-1)$ with h the multiplicity of the root p_m.

With inputs and outputs as defined by Equation (3-35), the effect of an individual input v_k on a particular output y_j is given by

$$y_j = \sum_{i=1}^{n} \int_{t_0}^{t} h_{ijk}\, e^{p_i(t-\tau)}\, v_k(\tau)\, d\tau + D_{jk}\, v_k + E_{jk}\, \dot{v}_k \tag{3-40}$$

where the h_{ijk} are constants readily computed from the K_{ijk} in Equation (3-38) and the elements of the matrices $\underset{\sim}{B}$ and $\underset{\sim}{C}$ in Equation (3-35).

Distinct p_i roots are assumed in Equation (3-40); otherwise there may also be terms in the form of (3-39).

In the special case of sinusoidal steady-state analysis, all the $v_k(t)$ have the form $V e^{j\omega t}$, where V is complex and ω is real. With such a forcing function, each term in the sum in Equation (3-40) becomes

$$\begin{aligned}\int_{t_0}^{t} h_{ijk}\, e^{p_i(t-\tau)}\, V\, e^{j\omega\tau}\, d\tau &= V\, h_{ijk}\, e^{p_i t} \int_{t_0}^{t} e^{(j\omega - p_i)\tau}\, d\tau \\ &= V\, h_{ijk}\, \frac{1}{j\omega - p_i} e^{p_i t} \left[e^{(j\omega - p_i)\tau}\right]_{\tau=t_0}^{\tau=t} \\ &= \frac{-V\, h_{ijk} e^{(j\omega - p_i)t_0}}{j\omega - p_i} e^{p_i t} + \frac{V h_{ijk}}{j\omega - p_i} e^{j\omega t}\end{aligned} \tag{3-41}$$

The first term in the last form of (3-41) is a transient of the same form as the transients due to initial conditions (the zero-input response). The second term is the steady-state response. If $v_k(t) = V\, e^{j\omega t}$ and all other sources are zero, then from Equation (3-40) the steady-state part of $y_j(t)$ due to v_k is

$$y_j(t) = V \left(\sum_{i=1}^{n} \frac{h_{ijk}}{j\omega - p_i} + D_{jk} + j\omega E_{jk} \right) e^{j\omega t} \tag{3-42}$$

From Equation (3-42) it is easy to relate the constants h_{ijk}, D_{jk}, and E_{jk} to the steady-state transfer functions $H(j\omega)$ that are often used in electrical engineering. We recall the procedure for computing steady-state transfer functions as follows:

1. Write a set of equilibrium integrodifferential equations for the system. (In many cases the normal form equations will be just as easy to write as any other set.)

2. Assume the input has the form $V\, e^{j\omega t}$, the output has the form $Y\, e^{j\omega t}$, and all other variables $w_i(t)$ that appear in the equilibrium equations have the form $W_i e^{j\omega t}$.

3. Note that differentiation of a variable is equivalent to multiplication of the assumed form by $j\omega$ and integration to division by $j\omega$. Furthermore, all terms in the equilibrium equations have the common factor $e^{j\omega t}$ which cancels out.

4. Replace the equilibrium differential equations by algebraic equations with Y and the W_i as unknowns, polynomials in $j\omega$ and $1/j\omega$ as the coefficients of the unknowns, and V as the known inhomogeneous part. This is the first step that one actually writes on paper.

5. Solve the algebraic equations for Y. The solution will have the form

$$Y = H(j\omega)\, V$$

Then the time function output is

$$y(t) = H(j\omega)\, V\, e^{j\omega t} \tag{3-43}$$

Relation between the *A* matrix and system poles. Since the coefficients of the equilibrium algebraic equations are polynomials in $(j\omega)$ or $(1/j\omega)$, the transfer function $H(j\omega)$ will be a rational function of $j\omega$ (a ratio of polynomials). A partial fraction expansion of $H(j\omega)$ in Equation (3-43) leads to exactly the form given in (3-42). Since the partial fraction expansion is unique, as is the steady-state response, the coefficients for the general time response (3-40) can be found by first finding the transfer function and then expanding in partial fractions. There is no need to go through the entire variation of parameters.

We should note that the above derivation implies that the poles of the transfer functions are the eigenvalues of the $\underset{\sim}{A}$ matrix. Every pole must be an eigenvalue, but if one of the h_{ijk} turned out to be zero, there would be eigenvalues that did not appear as poles. If there is a double-order pole, there is a multiple eigenvalue. The rules for partial fraction expansion lead to the form (3-39) directly in the multiple eigenvalue case. Thus, the transfer function expansion gives the right answer for the multiple eigenvalue case as well as the distinct root case.

As an example consider the critically damped second-order system characterized by the differential equation,

$$\ddot{y} + 2\dot{y} + y = \dot{v} - v \tag{3-44}$$

where y is the output and v is the input.

In order to find the total response $y(t)$ for a given set of initial conditions $y(t_0)$ and $\dot{y}(t_0)$, we could proceed as in Section 2.4 and set up state variable equations. Then the method of Section 3.1 could be used to solve the equations. However, since we know the form of the answer, we need only solve Equation (3-44) in the sinusoidal steady state to get the constants in the forced response terms. Below we shall see how to get the natural response terms as well.

Specifically we assume

$$\left.\begin{aligned} y(t) &= Y\, e^{j\omega t} \\ v(t) &= V\, e^{j\omega t} \end{aligned}\right\} \tag{3-45}$$

Then Equation (3-44) becomes

$$((j\omega)^2 + 2j\omega + 1)\, Y = (j\omega - 1)\, V \tag{3-46}$$

The transfer function is

$$H(j\omega) = \frac{j\omega - 1}{(j\omega)^2 + 2j\omega + 1} = \frac{-2}{(j\omega + 1)^2} + \frac{1}{(j\omega + 1)} \tag{3-47}$$

Using the correspondence between the coefficients of the transfer function and the constants in the forced response, we have the forced response (zero-state) term for (3-44) as

$$y(t) = \int_{t_0}^{t} v(\tau)\ [e^{-(t-\tau)} - 2(t-\tau)\ e^{-(t-\tau)}]\ d\tau \tag{3-48}$$

The correctness of the second term in the bracket is readily verified by the steps of Equation (3-41).

Computation of the state transition matrix. The steady-state analysis method can also be used to compute the state transition matrix and the natural, or zero-input, response when the state vector is known at $t = t_0$. In every solution matrix each column is a solution to the homogeneous system of equations. The n columns form a set of n linearly independent solutions. Since any other solution can be written as a linear combination of the members of such a linear independent set of solutions (see Reference 11), each column of one fundamental matrix can be constructed as a linear combination of the columns of any other fundamental matrix. In matrix terms this means: Given any two fundamental matrices ϕ_a and ϕ_b, ϕ_b can be constructed from ϕ_a by

$$\underset{\sim}{\phi}_b = \underset{\sim}{\phi}_a \underset{\sim}{K} \tag{3-49}$$

where $\underset{\sim}{K}$ is a nonsingular matrix of constants. The matrix K must be nonsingular since both $\underset{\sim}{\phi}_a$ and $\underset{\sim}{\phi}_b$ must be nonsingular if they are solution matrices.

With the above fact about the relation between two solution matrices, let us examine the forced response term more closely. For given solution matrix $\underset{\sim}{\phi}_a$ the forced response contains the matrix product, $\underset{\sim}{\phi}_a(t)\ \underset{\sim}{\phi}_a^{-1}(\tau)$. For given $\underset{\sim}{\phi}_b$ the term contains

$$\underset{\sim}{\phi}_b(t)\ \underset{\sim}{\phi}_b^{-1}(\tau) = \underset{\sim}{\phi}_a(t)\ \underset{\sim}{K}\ \underset{\sim}{K}^{-1}\ \underset{\sim}{\phi}_a^{-1}(\tau) = \underset{\sim}{\phi}_a(t)\ \underset{\sim}{\phi}_a^{-1}(\tau) \tag{3-50}$$

Thus, the forced response term in Equation (3-16) is independent of the choice of fundamental matrix.

From the discussion about the functional form (3-37) above we see that for any fundamental matrix, the product matrix (3-50) is a function of the single variable $(t-\tau)$. Let

$$\underset{\sim}{\psi}(t-\tau) = \underset{\sim}{\phi}(t)\ \underset{\sim}{\phi}^{-1}(\tau) \tag{3-51}$$

When the state transition matrix is used in Equation (3-51), we have

$$\underset{\sim}{\psi}(t-\tau) = \underset{\sim}{\phi}(t, t_0)\ \underset{\sim}{\phi}^{-1}(\tau, t_0) \tag{3-52}$$

Since $\phi(t_0, t_0)$ is the identity matrix, evaluation of Equation (3-52) for τ equal to t_0 shows that

$$\underset{\sim}{\phi}(t, t_0) = \underset{\sim}{\psi}(t-t_0) \tag{3-53}$$

The elements of $\underset{\sim}{\psi}(t)$ are easily computed by steady-state analysis. Consider the algebraic equation

$$j\omega \mathbf{X} = \underset{\sim}{A} \mathbf{X} + \mathbf{U} \tag{3-54}$$

where $\mathbf{X}$ and $\mathbf{U}$ are vectors of complex numbers. $\underset{\sim}{A}$ is the matrix from Equation (3-1).

Regroup the equation and solve for the components of $\mathbf{X}$ in terms of the components of $\mathbf{U}$. Thus, in terms of determinants,

$$X_i = \sum_{j=1}^{n} -\frac{\Delta_{ji} U_j}{\Delta} \tag{3-55}$$

where Δ is the determinant

$$\begin{vmatrix} a_{11} - j\omega & a_{12} & \cdots & a_{1n} \\ a_{21} & a_{22} - j\omega & \cdots & a_{2n} \\ \vdots & \vdots & \vdots\vdots\vdots & \vdots \\ a_{n1} & a_{n2} & \cdots & a_{nn} - j\omega \end{vmatrix}$$

Δ_{ij} is the cofactor (with sign) of the *ij*th element of the matrix from which Δ is formed.

The transfer function $-(\Delta_{ij}/\Delta)$ is a rational function in $j\omega$. It can be expanded in partial fractions. Since Δ is a polynomial of degree n, and Δ_{ij} is of degree $(n-1)$, the expansion has no constant term. Clearly, the roots of the denominator are the eigenvalues of $\underset{\sim}{A}$. Now

$$-\frac{\Delta_{ji}}{\Delta} = \frac{h_{ij1}}{j\omega - p_i} + \frac{h_{ij2}}{j\omega - p_2} + \cdots + \frac{h_{ijn}}{j\omega - p_n} \tag{3-56}$$

Then

$$\phi_{ij}(t, t_0) = h_{ij1}\, e^{p_1(t-t_0)} + h_{ij2}\, e^{p_2(t-t_0)} + \cdots + h_{ijn}\, e^{p_n(t-t_0)} \tag{3-57}$$

The application of this procedure to the determination of the natural response of Equation (3-44) requires an examination of only the homogeneous part of that equation. An appropriate $\underset{\sim}{A}$ matrix for Equation (3-44) is

$$\underset{\sim}{A} = \begin{bmatrix} 0 & 1 \\ -1 & -2 \end{bmatrix} \tag{3-58}$$

The determinant Δ in Equation (3-55) is

$$\Delta = \begin{vmatrix} -j\omega & 1 \\ -1 & -2 - j\omega \end{vmatrix} = (j\omega)^2 + 2j\omega + 1$$

The cofactors are

$$\Delta_{11} = -2 - j\omega \qquad \Delta_{12} = 1$$
$$\Delta_{21} = -1 \qquad \Delta_{22} = -j\omega$$

The four normalized cofactors for Equation (3-51) are

$$-\frac{\Delta_{11}}{\Delta}=\frac{1}{(j\omega+1)^2}+\frac{1}{j\omega+1}$$

$$-\frac{\Delta_{12}}{\Delta}=\frac{-1}{(j\omega+1)^2}$$

$$-\frac{\Delta_{21}}{\Delta}=\frac{1}{(j\omega+1)^2}$$

$$-\frac{\Delta_{22}}{\Delta}=\frac{-1}{(j\omega+1)^2}+\frac{1}{j\omega+1}$$

By the correspondence between Equations (3-56) and (3-57), the state transition matrix is seen to be

$$\underset{\sim}{\phi}(t, t_0) = \begin{bmatrix} (t-t_0+1)\, e^{-(t-t_0)} & -1(t-t_0)\, e^{-(t-t_0)} \\ (t-t_0)\, e^{-(t-t_0)} & (1-t+t_0)\, e^{-(t-t_0)} \end{bmatrix} \tag{3-59}$$

With the designation of the A matrix (3-58) corresponding to Equation (3-44), the state variable x_1 is the output y. Thus, the natural response part of y is the zero-input part of x_1. It is

$$x_1(\text{zero input}) = x_1(t_0)\, \phi_{11}(t, t_0) + x_2(t_0)\, \phi_{12}(t, t_0) \tag{3-60}$$

But $x_1(t_0) = y(t_0)$ and $x_2(t_0) = \dot{y}(t_0)$. Thus, if $y(t_0)$ and $\dot{y}(t_0)$ are given along with $v(t)$ for t between t_0 and t, the total response of Equation (3-44) is found by adding (3-60) to (3-48). The result is

$$y(t) = \int_{t_0}^{t} v(\tau)(1-2t+2\tau)\, e^{-(t-\tau)}\, d\tau + [y(t_0)(t-t_0+1) + \dot{y}(t_0)(t-t_0)]\, e^{-(t-t_0)} \tag{3-61}$$

In addition to the above method, which is most convenient for hand computation, there is a series method for obtaining the state transition matrix that is useful for digital computer application. The exponential function has a power series representation

$$e^{pt} = 1 + pt + \frac{(pt)^2}{2\,!} + \cdots \tag{3-62}$$

One can readily show that this power series satisfies the first-order homogeneous differential equation

$$\dot{x} = px \tag{3-63}$$

By the same reasoning, but with matrix multiplication, the matrix equation (3-10) has a power series solution

$$\underset{\sim}{\phi}(t, t_0) = \underset{\sim}{I} + (t-t_0)\, \underset{\sim}{A} + \frac{(t-t_0)^2}{2\,!}\underset{\sim}{A}\, \underset{\sim}{A} + \cdots \tag{3-64}$$

The power series (3-64) as a matrix has many properties of the exponential function. Thus, a common notation is

$$\phi(t, t_0) = e^{\underset{\sim}{A}(t-t_0)} \tag{3-65}$$

One can readily show that when $\underset{\sim}{\phi}(t, t_0)$ is given by Equation (3-64)

$$e^{\underset{\sim}{A} t_1} e^{\underset{\sim}{A} t_2} = e^{\underset{\sim}{A}(t_1+t_2)} \tag{3-66}$$

$$\phi^{-1}(t, t_0) = e^{-\underset{\sim}{A}(t-t_0)} \tag{3-67}$$

$$\phi(t, t_0)\, \phi^{-1}(\tau, t_0) = e^{\underset{\sim}{A}(t-\tau)} \tag{3-68}$$

When the matrix multiplications of Equation (3-64) are carried out, each element of the resulting matrix is a power series in the form

$$\phi_{ij}(t) = b_0 + b_1 t + b_2 \frac{t^2}{2\,!} + b_3 \frac{t^3}{3\,!} + \cdots \tag{3-69}$$

Each such term can be regrouped as a sum of power series expansions of the n exponential functions with the $p_i t$ as exponents. When there are multiple p_i values, the appropriate $t\, e^{p_i t}$ terms occur accordingly.

For the $\underset{\sim}{A}$ matrix (3-58)

$$\underset{\sim}{A}^2 = \begin{bmatrix} 0 & 1 \\ -1 & -2 \end{bmatrix} \begin{bmatrix} 0 & 1 \\ -1 & -2 \end{bmatrix} = \begin{bmatrix} -1 & -2 \\ 2 & 3 \end{bmatrix}$$

$$\underset{\sim}{A}^3 = \begin{bmatrix} -1 & -2 \\ 2 & 3 \end{bmatrix} \begin{bmatrix} 0 & 1 \\ -1 & -2 \end{bmatrix} = \begin{bmatrix} 2 & 3 \\ -3 & -4 \end{bmatrix}$$

Thus, the first four terms in the power series (3-64) for this example are

$$\underset{\sim}{\phi}(t, t_0) =$$

$$\begin{bmatrix} 1 - \dfrac{(t-t_0)^2}{2} + \dfrac{(t-t_0)^3}{3} + \cdots & (t-t_0) - (t-t_0)^2 + \dfrac{(t-t_0)^3}{2} + \cdots \\ -(t-t_0) + (t-t_0)^2 - \dfrac{(t-t_0)^3}{2} + \cdots & \\ & 1 - 2(t-t_0) + \dfrac{3(t-t_0)^2}{2} - \dfrac{2(t-t_0)^3}{3} + \cdots \end{bmatrix} \tag{3-70}$$

As a check let us expand the terms in Equation (3-59) in power series and see if they are the same as those of Equation (3-70). From (3-59)

$$\phi_{11}(t-t_0) = (t - t_0 + 1)\left(1 - (t-t_0) + \frac{(t-t_0)^2}{2} - \frac{(t-t_0)^3}{6} + \cdots\right)$$

$$= 1 - \frac{(t-t_0)^2}{2} + \frac{(t-t_0)^3}{3} + \text{terms of order } (t-t_0)^4$$

$$\phi_{12}(t-t_0) = (t-t_0) - (t-t_0)^2 + \frac{(t-t_0)^3}{2} + \cdots$$

$$\phi_{21}(t-t_0) = -\phi_{12}(t-t_0)$$

$$\phi_{22}(t-t_0) = (1-(t-t_0))\left(1-(t-t_0)+\frac{(t-t_0)^2}{2}-\frac{(t-t_0)^3}{6}+\cdots\right)$$

$$= 1 - 2(t-t_0) + \frac{3(t-t_0)^2}{2} - \frac{2(t-t_0)^3}{3} + \cdots$$

Thus, all terms check.

In this section we have shown that all the time-domain quantities derived by variation of parameters in Section 3-1 can be computed much more easily by frequency-domain methods. One might question why the awkward $j\omega$ notation was used rather than the neater Laplace transform notation. In Laplace transform solution to differential equations the time origin must be set. If the initial time t_0 is not taken at zero, the notation becomes very messy. Furthermore, if the initial time is allowed to approach minus infinity, as is convenient in many steady-state problems of communications, control, and circuit theory, there are serious questions of convergence of the improper transform integral. For this reason the $j\omega$ notation was used to remind the reader that the manipulations are for computational purposes only. The truth of the results is so far based on the variation of parameters method and not on formal transform techniques.

PROBLEMS

3-1 Consider the third-order system with $\underset{\sim}{A}$ matrix $\begin{bmatrix} -2 & 1 & 0 \\ 0 & -2 & 1 \\ 0 & 0 & -2 \end{bmatrix}$.

Since the natural frequencies are all at -2, three linearly independent solutions to the homogeneous equation are e^{-2t}; te^{-2t}; and t^2e^{-2t}. Find the state transition matrix $\underset{\sim}{\Phi}(0, t)$ using the following facts:

a. The columns of the matrix satisfy the homogeneous equation.

b. Each solution x_i to the homogeneous equation has the form

$$B_{i1}e^{-2t} + B_{i2}te^{-2t} + B_{i3}t^2e^{-2t}$$

c. At $t = 0$, $\underset{\sim}{\Phi}(0, 0)$ is the 3×3 identity matrix.

3-2 Consider the third-order system with $\underset{\sim}{A}$ matrix $\begin{bmatrix} -2 & 1 & 0 \\ 0 & -2 & 0 \\ 0 & 1 & -2 \end{bmatrix}$.

As in Problem 3-1 the natural frequencies are all at -2. Use the procedure outlined in that problem to find the state transition matrix. How does this system differ from the system of Problem 3-1?

3-3 Consider the first-order, time-variant system of Figure (3-1) with $R(t) = R_0 + R_1 \cos \omega_0 t$ and $v(t) = V \cos (\omega_1 t + \theta)$; $R_0, R_1, V, \omega_0, \omega_1, \theta$ all real constants. Let $i(t_0) = 0$. Compute the steady-state response at time t by taking the limit of Equation (3-31) as $t_0 \to -\infty$.

Often in the engineering literature one sees a frozen system function used for time-variant systems. In such cases an approximate steady-state response is computed by first analyzing the system as if all its parameters were fixed. Then in the final answer the variable parameters are substituted for the fixed parameters. Thus, for the circuit of Figure 3-1, the approximate steady-state response would be

$$i_{\text{approx}}(t) = \frac{V \cos (\omega_1 t + \theta + \alpha(t))}{[(R_0 + R_1 \cos \omega_0 t)^2 + \omega_1^2 L^2]^{1/2}}$$

where

$$\alpha(t) = \tan^{-1} \frac{\omega_1 L}{R_0 + R_1 \cos \omega_0 t}$$

Show the error in this approximation.

Hint: Write i_{approx} in phasor notation and note

$$\frac{e^{j\omega_1 t}}{R_0 + R_1 \cos \omega_0 t + j\omega_1 L}$$

$$= \frac{\exp\left(\frac{R_0}{L} + \frac{R_1}{\omega_0 L} \sin \omega_0 t\right)}{L} \int_{-\infty}^{t} \frac{d}{d\tau}\left\{\frac{\exp\left[-\left(\frac{R_0}{L} + j\omega_1\right)\tau - \frac{R_1}{\omega_0 L} \sin \omega_0 \tau\right]}{\frac{R_0}{L} + j\omega_1 + \frac{R_1}{L} \cos \omega_0 \tau}\right\} d\tau$$

3-4 A rotor of moment of inertia one is rolling on frictionless bearings with angular velocity one. At $t = 0$, oil begins to leak into one bearing and the friction coefficient starts increasing linearly to one at the end of 1 second. Then the leak stops and the frictional coefficient remains constant at one. Meanwhile, at $t = 0$, a triangular torque pulse is started. The torque increases linearly from zero to one in 1 second and then decreases to zero linearly during the next second.

Write an expression for the angular velocity of the shaft for all time. Do not try to integrate the messy expressions. Leave them as integrals that could be performed by anyone who has completed freshman calculus and has a good set of tables.

3-5 Set up an analog computer diagram for the system

$$\dot{\mathbf{x}} = \begin{bmatrix} -1 & 1 & -2 \\ 0 & -2 & 1 \\ 0 & -2 & -2 \end{bmatrix} \mathbf{x} + \begin{bmatrix} 1 & 1 \\ 0 & 1 \\ 1 & 0 \end{bmatrix} \mathbf{v}$$

$$y = [1 \quad 0 \quad 3]\, \mathbf{x} + [2 \quad 0]\, \mathbf{v} + [1 \quad 0]\, \dot{\mathbf{v}}$$

3-6 The circuit of Figure P3-6 is a 3-port. Write state variable equations with input and output in the form of Equation (3-35) wherein the currents at the ports are inputs and the voltages outputs.

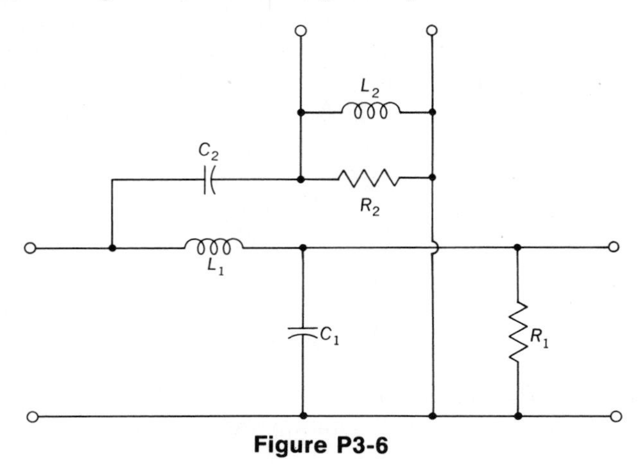

Figure P3-6

3-7 Using the steady-state analysis method of Section 3-3, find the state transition matrix of the system whose $\underset{\sim}{A}$ matrix is $\begin{bmatrix} -2 & 1 & 0 \\ 0 & -2 & 1 \\ 0 & 0 & -2 \end{bmatrix}$.

Compare this answer with the result of Problem 3-1.

3-8 Repeat Problem 3-2 using the method of Problem 3-6.

3-9 Set up normal form equations plus an output equation in the form (3-35) for the 2-port circuit whose impedance matrix is

$$\begin{bmatrix} \dfrac{1}{\sqrt{2}} \left[\dfrac{j\omega + \sqrt{2}}{-(j\omega)^2 + \sqrt{2}\, j\omega + 1} \right] & \dfrac{1}{(j\omega)^2 + \sqrt{2}\, j\omega + 1} \\ \dfrac{1}{(j\omega)^2 + \sqrt{2}\, j\omega + 1} & \dfrac{(j\omega)^2 + 1}{(j\omega)^2 + \sqrt{2}\, j\omega + 1} \end{bmatrix}$$

Recall that a circuit with given impedance matrix has inputs as currents and outputs as voltages.

CHAPTER

4

General Linear Systems Described by the Convolution Integral

In the previous chapter, the zero-state response of a lumped, linear, time-invariant system was shown to be the convolution of exponential functions (or the product of polynomials with exponential functions) with the input. In this chapter the convolution integral is discussed from a more general point of view. The discussion shows that any system whose zero-state response is a convolution of input and system function is linear and time-invariant. This fact, along with certain algebraic properties of convolution, leads to a discussion of block diagrams as a means of representing linear systems. In the final section of this chapter, these general block diagrams are related to the elemental diagrams and $\underset{\sim}{A}$ matrix characterization of Chapter 2.

4-1 Properties of Convolutions

The basic convolution form that arises in the solution to the equations of the previous chapter [for example, Equation (3-38)] is

$$\int_{t_0}^{t} f(t-\tau)\, g(\tau)\, d\tau \tag{4-1}$$

The specific correspondence between Equations (3-38) and (4-1) is

$$g(t) = u_j(t)$$

$$f(t) = \sum_{k=1}^{n} K_{ijk}\, e^{p_k t}$$

Then (3-38) is a sum of n terms of the form (4-1). Later in this section we shall generalize the function $f(t)$ so that the entire right side of (3-40) can be considered in the form of (4-1).

Algebraic properties. Convolution integrals have some interesting mathematical properties that can be exploited for useful engineering applications of linear systems. One such property is the similarities between the convolution of two functions and the multiplication of two numbers. Another is the form of the result of differentiating or integrating a convolution.

The multiplication of two quantities is an important operation relative to the addition of those quantities. For integrable functions such as f and g in (4-1), addition is readily defined as

$$(f+g)\ (t) = f(t) + g(t) \tag{4-2}$$

That is, the sum of two functions is obtained by adding the members of the range corresponding to a specific member of the domain. Viewed as a second operation between functions, convolution is distributive with respect to addition. That is, if we define a function h by the sum of two functions f and g the convolution of u with h is equal to the convolution of u with f, plus the convolution of u with g. In symbols,

$$\begin{gathered} h = f + g \\ \int_{t_0}^{t} u(t-\tau)\, h(\tau)\, d\tau = \int_{t_0}^{t} u(t-\tau)\, f(\tau)\, d\tau + \int_{t_0}^{t} u(t-\tau)\, g(\tau)\, d\tau \end{gathered} \tag{4-3}$$

In the special case when $t_0 = 0$, the convolution of two functions is also commutative. That is,

$$\int_{0}^{t} f(t-\tau)\, g(\tau)\, d\tau = \int_{0}^{t} g(t-\tau)\, f(\tau)\, d\tau \tag{4-4}$$

To prove Equation (4-4), we make the change of variables $v = t - \tau$ in Equation (4-1). Thus,

$$\int_{t_0}^{t} f(t-\tau)\, g(\tau)\, d\tau = \int_{0}^{t-t_0} f(v)\, g(t-\nu)\, d\nu \tag{4-5}$$

In the special case when $t_0 = 0$, Equation (4-5) becomes Equation (4-4).

To put specific numbers into the formulas (4-3) and (4-4), let

$$f(t) = \begin{cases} 1 & \text{for } 0 \leqslant t \leqslant 2 \\ 0 & \text{otherwise} \end{cases} \tag{4-6}$$

$$g(t) = \begin{cases} 1 & \text{for } 1 \leqslant t \leqslant 3 \\ 0 & \text{otherwise} \end{cases} \tag{4-7}$$

$$u(t) = e^{-t} \quad \text{for all } t \tag{4-8}$$

Then $h(t)$ in (4-3) is

$$h(t) = \begin{cases} 1 & \text{for } 0 \leqslant t < 1 \\ 2 & \text{for } 1 \leqslant t \leqslant 2 \\ 1 & \text{for } 2 < t \leqslant 3 \\ 0 & \text{otherwise} \end{cases} \tag{4-9}$$

Assuming that $t_0 \leqslant 0$, the left side of Equation (4-3) becomes:

for $0 \leqslant t \leqslant 1$

$$\int_0^t e^{-(t-\tau)}\, d\tau = 1 - e^{-t} \tag{4-10a}$$

for $1 \leqslant t \leqslant 2$

$$\int_0^1 e^{-(t-\tau)}\, d\tau + \int_1^t 2e^{-(t-\tau)}\, d\tau = e^{-t}\,(e-1) + 2(1 - e^{-(t-1)}) \tag{4-10b}$$

for $2 \leqslant t \leqslant 3$

$$\begin{aligned} &\int_0^1 e^{-(t-\tau)}\, d\tau + \int_1^2 2e^{-(t-\tau)}\, d\tau + \int_2^t e^{-(t-\tau)}\, d\tau \\ &\qquad = e^{-t}(e-1) + 2e^{-t}(e^2 - e) + 1 - e^{-(t-2)} \end{aligned} \tag{4-10c}$$

for $3 \leqslant t$

$$\begin{aligned} \int_0^t h(t-\tau)\, e^{-\tau} d\tau &= e^{-t}\,[e - 1 + 2(e^2 - e) + e^3 - e^2] \\ &= e^{-t}\,[e^3 + e^2 - e - 1] \end{aligned} \tag{4-10d}$$

To compare with the right side of Equation (4-3) we compute

$$\int_0^t u(t-\tau)\, f(\tau)\, d\tau = \begin{cases} 1 - e^{-t} & \text{for } 0 \leqslant t \leqslant 2 \\ e^{-t}\,[e^2 - 1] & \text{for } t \geqslant 2 \end{cases} \tag{4-11a}$$

and

$$\int_0^t u(t-\tau)\, g(\tau)\, d\tau = \begin{cases} 0 & \text{for } 0 < t \leqslant 1 \\ 1 - e^{-(t-1)} & \text{for } 1 \leqslant t \leqslant 3 \\ e^{-t}\,(e^3 - e) & \text{for } t \geqslant 3 \end{cases} \tag{4-11b}$$

Combining Equations (4-11a) and (4-11b) gives

$$\int_{t_0}^{t} u(t-\tau)(f+g)(\tau)\, d\tau = \begin{cases} 1-e^{-t} & \text{for } 0 \leqslant t \leqslant 1 \\ 2-e^{-t}(1+e) & \text{for } 1 \leqslant t \leqslant 2 \\ 1+e^{-t}(e^2-e-1) & \text{for } 2 \leqslant t \leqslant 3 \\ e^{-t}(e^3+e^2-e-1) & \text{for } 3 \leqslant t \end{cases} \tag{4-12}$$

The four parts of Equation (4-12) are exactly the same as Equations (4-10a), (4-10b), (4-10c), and (4-10d), respectively.

To demonstrate the computations associated with the commutative formula (4-4), let us convolve f from Equation (4-6) and g from Equation (4-7) both ways. First:

$$\int_0^t f(t-\tau)\, g(\tau)\, d\tau = \begin{cases} 0 & \text{for } t \leqslant 1 \\ \displaystyle\int_1^t d\tau = t-1 & \text{for } 1 \leqslant t \leqslant 3 \\ \displaystyle\int_1^3 f(t-\tau)\, d\tau = \begin{cases} 5-t \\ 0 \end{cases} & \begin{matrix} \text{for } 3 \leqslant t \leqslant 5 \\ \text{for } 5 \leqslant t \end{matrix} \end{cases} \tag{4-13}$$

Then

$$\int_0^t g(t-\tau)\, f(\tau)\, d\tau = \begin{cases} \displaystyle\int_0^t g(t-\tau)\, d\tau & \text{for } 0 \leqslant t \leqslant 2 \\ \displaystyle\int_0^2 g(t-\tau)\, d\tau & \text{for } t \geqslant 2 \end{cases} \tag{4-14}$$

But

$$\int_0^t g(t-\tau)\, d\tau = \begin{cases} 0 & \text{for } t \leqslant 1 \\ (t-1) & \text{for } 1 \leqslant t \leqslant 3 \\ 2 & \text{for } 3 \leqslant t \end{cases} \tag{4-15a}$$

and

$$\int_0^2 g(t-\tau)\, d\tau = \begin{cases} 0 & \text{for } t \leqslant 1 \\ (t-1) & \text{for } 1 \leqslant t \leqslant 3 \\ (5-t) & \text{for } 3 \leqslant t \leqslant 5 \\ 0 & \text{for } 5 \leqslant t \end{cases} \tag{4-15b}$$

Combining Equations (4-15a) and (4-15b) with Equation (4-14) gives exactly Equation (4-13).

Differentiation and integration of convolutions. When an output signal that is a convolution of two functions is used as an input to a subsequent system, the operations of integration and differentiation of convolutions becomes important. Let f and g be continuously differentiable, and let

$$h(t) = \int_{t_0}^{t} f(t-\tau)\, g(\tau)\, d\tau \tag{4-16}$$

By Leibniz formula for differentiating integrals (see Reference 24, p. 262),

$$h(t) = f(0)\, g(t) + \int_{t_0}^{t} \dot{f}(t-\tau)\, g(\tau)\, d\tau \tag{4-17}$$

In Equation (4-17) the dot over the function in the integrand means the derivative of f with respect to its argument. Thus, if

$$f(t) = \cos t$$

$$\dot{f}(t) = -\sin t$$

and

$$\dot{f}(t-\tau) = -\sin\,(t-\tau)$$

Using the formula (4-5) before applying Leibniz formula gives the alternate form

$$\dot{h}(t) = g(t_0)\, f(t-t_0) + \int_{0}^{t-t_0} f(\tau)\, \dot{g}(t-\tau)\, d\tau \tag{4-18}$$

As derived, the formulas (4-17) and (4-18) apply only when f and g are continuously differentiable. With only minor modification the formulas also apply when f and g have jump discontinuities at discrete points. Signals with such jumps are very common in engineering. Suppose f has a jump discontinuity at t_1 but is otherwise continuously differentiable. For definiteness let

$$\lim_{\substack{t \to t_1 \\ t < t_1}} f(t) = f(t_1) \qquad \lim_{\substack{t \to t_1 \\ t > t_1}} f(t) = f(t_1) + \alpha$$

Then

$$h(t) = \int_{t_0}^{t-t_1} f(t-\tau)\, g(\tau)\, d\tau + \int_{t-t_1}^{t} f(t-\tau)\, g(\tau)\, d\tau$$

Differentiating gives

$$\dot{h}(t) = \int_{t_0}^{t-t_1} \dot{f}(t-\tau)\, g(\tau)\, d\tau + [f(t_1) + \alpha]\, g(t-t_1)$$

$$+ \int_{t-t_1}^{t} \dot{f}(t-\tau)\, g(\tau)\, d\tau - f(t_1)\, g(t-t_1) + f(0)\, g(t)$$

or

$$\dot{h}(t) = f(0)\, g(t) + \alpha\, g(t-t_1) + \int_{t_0}^{t} \dot{f}(t-\tau)\, g(\tau)\, d\tau \tag{4-19}$$

where $\dot{f}$ may have jump discontinuities. A similar formula for $\dot{h}(t)$ from Equation (4-16) can also be derived when g has jump discontinuities.

A formula similar to Equation (4-17) for the integral of $h(t)$ of Equation (4-16) is readily derived by integration by parts. Thus,

$$\int_{t_0}^{t} h(\tau)\,d\tau = \int_{t_0}^{t}\int_{t_0}^{\tau} f(\tau-\nu)\,g(\nu)\,d\nu\,d\tau$$

But

$$\int_{t_0}^{\tau} f(\tau-\nu)\,g(\nu)\,d\nu = \left[f(\tau-\nu)\int_{t_0}^{\nu} g(\lambda)\,d\lambda\right]_{\nu=t_0}^{\nu=\tau} - \int_{t_0}^{\tau} -\dot{f}(\tau-\nu)\int_{t_0}^{\nu} g(\lambda)\,d\lambda\,d\nu$$

or

$$\int_{t_0}^{\tau} f(\tau-\nu)\,g(\nu)\,d\nu = f(0)\int_{t_0}^{\tau} g(\lambda)\,d\lambda + \int_{t_0}^{\tau} \dot{f}(\tau-\nu)\int_{t_0}^{\nu} g(\lambda)\,d\lambda\,d\nu \tag{4-20}$$

But by Equation (4-17), the right side of Equation (4-19) is just

$$\frac{d}{d\tau}\left[\int_{t_0}^{\tau} f(\tau-\nu)\int_{t_0}^{\nu} g(\lambda)\,d\lambda\,d\nu\right]$$

Thus,

$$\int_{t_0}^{t} h(\tau)\,d\tau = \int_{t_0}^{t} f(t-\nu)\int_{t_0}^{\nu} g(\lambda)\,d\lambda\,d\nu \tag{4-21}$$

If the derivation of Equation (4-21) started with the change of variables (4-5), then the resulting formula would have been

$$\int_{t_0}^{t} h(\tau)\,d\tau = \int_{0}^{t-t_0} g(t-\nu)\int_{t_0}^{\nu} f(\lambda)\,d\lambda\,d\nu \tag{4-22}$$

Convolutions with infinite limits. The formulas for the integral, (4-21) and (4-22), and those for the derivative, (4-16) and (4-17), are symmetric in f and g when $t_0 = 0$. This symmetry plus the distributive and commutative properties of convolutions shown in (4-3), (4-4), and (4-5) makes it easy to analyze system interconnections wherein the input-output relation of each component is a convolution. When the above formulas are used to develop the interconnection rules, all initial conditions must be taken at $t = 0$ if the results are to be simple enough for easy use. In many engineering problems it is just as easy to take initial conditions at $t_0 = 0$ as at any other time. In other problems we would essentially like steady-state results; that is, results when $t_0 \to -\infty$. In such cases it is not convenient to make $t_0 = 0$ the basic starting time. A generalization of the idea of convolution that includes Equation (4-1) as a special case is

$$\int_{-\infty}^{\infty} f(t-\tau)\, g(\tau)\, d\tau \tag{4-23}$$

As shown below this general form has all the nice properties of the $(0, t)$ convolution and it can handle any starting time including $t_0 = 0$ and $t_0 = -\infty$.

Because of the infinite limits on (4-23) there is a question of existence of the integral that is not trivial. To carry out a complete, rigorous discussion of the formula (4-23) requires the theory of distributions. (See Reference 29 for a treatment of the subject at the level of mathematical maturity of engineering graduate students.) For the present discussion many fine points that must be considered in nonlinear systems will be glossed over.

For the formula (4-23) to be meaningful, both f and g must be defined for all t from $-\infty$ to ∞. If $f(t)$ is zero for $t < 0$ and $g(t)$ is zero for $t < t_0$, then (4-23) and (4-1) are identical for $t > t_0$. Furthermore, both f and g must be well enough behaved as $t \to \pm\infty$ for the integrals to exist.[1] With the change of variables that lead to Equation (4-5), the formula (4-23) becomes

$$\begin{aligned} \int_{-\infty}^{\infty} f(t-\tau)\, g(\tau)\, d\tau &= \int_{\infty}^{-\infty} f(v)\, g(t-v)\,(-dv) \\ &= \int_{-\infty}^{\infty} g(t-\tau)\, f(\tau)\, d\tau \end{aligned} \tag{4-24}$$

Thus, just as in the $(0, t)$ convolution, the $(-\infty, \infty)$ convolution is commutative. To simplify the notation, we define

$$f * g(t) = \int_{-\infty}^{\infty} f(t-\tau)\, g(\tau)\, d\tau \tag{4-25}$$

The two operations $+$ defined by Equation (4-2) and $*$ defined by (4-25) between functions have all the properties of addition and multiplication of numbers. Subtraction is also well-defined for functions; but division presents some problems. These will be discussed as they arise.[2]

Differentiation and integration of the $(-\infty, \infty)$ convolution are particularly easy since the limits on the integral are definite and t appears only in the argument of one function. (If the differentiated function and its derivative are not both zero at $\pm\infty$, then the limits cannot be dispensed with so easily.) Thus, if

---

[1] The question of existence in the usual sense of numbers versus existence in the sense of distributions is another question beyond this text.

[2] See Problem 4-4 for some basic ideas on division of functions.

$$h(t) = \int_{-\infty}^{\infty} f(t-\tau)\, g(\tau)\, d\tau = f * g(t) \tag{4-26}$$

then

$$\dot{h}(t) = \int_{-\infty}^{\infty} \dot{f}(t-\tau)\, g(\tau)\, d\tau = \dot{f} * g(t)$$

Furthermore, by Equation (4-24),

$$\dot{h}(t) = f * \dot{g}(t) = \dot{f} * g(t) \tag{4-27}$$

The $(-\infty, \infty)$ convolutions require that at least one of the functions goes to zero sufficiently fast at both $\pm\infty$ so that the integral is well-defined. For any function $h(t)$ that goes to zero sufficiently fast at $-\infty$ so that $\int_{-\infty}^{t} h(\tau)\, d\tau$ exists, we can consider this integral as a signal. In this case differentiation and integration are inverse operations. Thus, for notational purposes, we define a dot under a symbol as

$$\underset{\cdot}{h}(t) = \int_{-\infty}^{t} h(\tau)\, d\tau \tag{4-28}$$

When $h(t)$ is given by Equation (4-26), an integration by parts gives

$$\underset{\cdot}{h}(t) = \underset{\cdot}{f} * g(t) = f * \underset{\cdot}{g}(t) \tag{4-29}$$

provided both f and g exist.

Since $\dot{\underset{\cdot}{h}}(t) = h(t)$,

$$h(t) = \dot{f} * \underset{\cdot}{g}(t) \tag{4-30}$$

Carrying this idea on we see that h with n dots above can be written infinitely many ways so long as the number of dots above f and g minus the number below is equal to the number of dots above h. A corresponding statement goes for dots below. In all of the above discussion the derivatives must exist and be continuous, and the appropriate functions must be well enough behaved at $\pm\infty$ for the integrals to exist. The techniques below handle certain functions whose derivatives do not exist in the normal sense. Still there are certain functions that are well-behaved, but whose successive integrals and/or derivatives are not. One such example is

$$f(t) = \begin{cases} \dfrac{1}{t} & \text{for } t \leqslant -1 \\ -\dfrac{5}{3} + \dfrac{2}{3}(-t)^{3/2} & \text{for } -1 \leqslant t \leqslant 0 \\ -\dfrac{5}{3} + \dfrac{2}{3}t^{3/2} & \text{for } 0 \leqslant t \leqslant 1 \\ -\dfrac{1}{t} & \text{for } 1 \leqslant t \end{cases} \tag{4-31}$$

Clearly, f does not exist since $\int_{-\infty}^{t} (1/\tau)\, d\tau$ does not exist. The derivative $\dot{f}$ exists and is continuous. It is

$$\dot{f}(t) = \begin{cases} -\dfrac{1}{t^2} & \text{for } t \leqslant -1 \\ -\sqrt{-t} & \text{for } -1 \leqslant t \leqslant 0 \\ \sqrt{t} & \text{for } 0 \leqslant t \leqslant 1 \\ \dfrac{1}{t^2} & \text{for } 1 \leqslant t \end{cases} \tag{4-32}$$

The second derivative $\ddot{f}$ does not exist at $t = \pm 1$ and at $t = 0$. At $t = \pm 1$ there are jump discontinuities in $\ddot{f}$. These can be taken care of by the methods described below. At the origin the problem is more serious for

$$\lim_{t \to 0} \ddot{f}(t) = \infty$$

whether $t \to 0$ from above or below. Within the scope of this text there is no general way to handle such functions as part of a convolution integrand.

Convolutions with discontinuous functions. By the technique used to derive (4-19) the differentiation of convolutions can be extended to include functions that are not continuously differentiable at discrete points. Thus, if $\dot{f}$ is well-defined except at t_1, and f has a jump discontinuity at t_1 so that $f(t_1 +) - f(t_1 -) = \alpha$, the derivation of Equation (4-19) gives

$$\dot{h}(t) = \dot{f}_0 * g(t) + \alpha\, g(t - t_1) \tag{4-33}$$

where f_0 is the continuous function

$$f_0(t) = f(t) - \alpha\, U(t - t_1)$$

with $U(t)$ the unit step function $- U(t) = \begin{cases} 0 & \text{for } t < 0 \\ 1 & \text{for } t \geqslant 0 \end{cases}$

Since f_0 is continuous and piecewise continuously differentiable, $\dot{f}_0$ has only jump discontinuities. The convolution $\dot{f}_0 * g(t)$ can be differentiated again by the formula (4-33). Then

$$\ddot{h}(t) = \overbrace{\dot{f}_0 * g(t)}^{\cdot} + \alpha\, \dot{g}(t - t) \tag{4-34}$$

This process can be carried to higher derivatives so long as $f(t)$ is suitably differentiable except at discrete points, and f and all needed derivatives go to zero at $\pm\infty$.

Formula (4-34) and its extension to higher derivatives is readily remembered by the use of Dirac delta function notation. The delta function is defined by

$$\int_{-\infty}^{\infty} g(t-\tau)\, \delta(\tau - t_0)\, d\tau = g(t-t_0) \tag{4-35}$$

Its successive derivatives are defined by

$$\int_{-\infty}^{\infty} g(t-\tau)\, \dot{\delta}(\tau - t_0)\, d\tau = \dot{g}(t-t_0) \tag{4-36}$$

and

$$\int_{-\infty}^{\infty} g(t-\tau)\, D^n\, \delta(\tau - t_0)\, d\tau = D^n\, g(t-t_0) \tag{4-37}$$

where D^n means the nth derivative.

The delta function is not a function. In this text it is merely a notational device, and has meaning only when appearing as an element of a convolution integral.[3] Symbolically, "the impulse is the derivative of a step function," or

$$\dot{U}(t) = \delta(t)$$

As a signal, the delta function is useful as something that delivers a finite amount of energy in zero time. In this context the word impulse function is often used. We shall make little use of this physical concept.

With the added notation of the delta function, a function that is piecewise continuously differentiable and has a finite number of jump discontinuities has a well-defined derivative when it appears in a convolution. Formulas (4-27), (4-28), (4-29) and their extensions apply so long as the functions in the convolutions are well-defined at all except a discrete set of points, and if delta functions are inserted to take care of those points. As an example of the use of these δ function techniques in formula (4-29), let us consider

$$f(t) = \begin{cases} t^2 & \text{for } 0 \leq t \leq 1 \\ 2-t & \text{for } 1 \leq t \leq 2 \\ 0 & \text{otherwise} \end{cases} \tag{4-38}$$

$$g(t) = \begin{cases} 0 & \text{for } t \leq 0 \\ e^{-t} & \text{for } 0 \leq t \end{cases} \tag{4-39}$$

Now

$$\dot{f}(t) = \begin{cases} 2t & \text{for } 0 \leq t \leq 1 \\ -1 & \text{for } 1 < t < 2 \\ 0 & \text{for } t < 0 \text{ and } t > 2 \end{cases} \tag{4-40}$$

[3] In distribution theory the delta function is a well-defined distribution and can stand alone.

Since $\dot{f}$ has jump discontinuities, $\ddot{f}$ will contain impulses. Thus,

$$\ddot{f}(t) = 2[U(t) - U(t-1)] - 3\delta(t-1) + \delta(t-2) \tag{4-41}$$

Continuing one step farther gives

$$\dddot{f}(t) = 2\delta(t) - 2\delta(t-1) - 3\dot{\delta}(t-1) + \dot{\delta}(t-2) \tag{4-42}$$

Since convolutions with δ functions and their derivatives are particularly easy to perform by use of formulas (4-34), (4-35), and (4-36), we can compute $h(t)$ by convolving $\dddot{f}$ with $\underset{\cdot\cdot\cdot}{g}$. First we need successive integrals of g. They are

$$\underset{\cdot}{g}(t) = \int_{-\infty}^{t} g(\tau)\, d\tau = \begin{cases} 0 & \text{for } t \leqslant 0 \\ \int_0^t e^{-\tau}\, d\tau & \text{for } t \geqslant 0 \end{cases}$$

$$= \begin{cases} 0 & \text{for } t \leqslant 0 \\ 1 - e^{-t} & \text{for } t \geqslant 0 \end{cases} \tag{4-43}$$

$$\underset{\cdot\cdot}{g}(t) = \begin{cases} 0 & \text{for } t \leqslant 0 \\ e^{-t} - 1 + t & \text{for } t \geqslant 0 \end{cases} \tag{4-44}$$

$$\underset{\cdot\cdot\cdot}{g}(t) = \begin{cases} 0 & \text{for } t \leqslant 0 \\ 1 - e^{-t} - t + \dfrac{t^2}{2} & \text{for } t \geqslant 0 \end{cases} \tag{4-45}$$

Combining the various formulas given:

$$h(t) = 2\,\underset{\cdot\cdot\cdot}{g}(t) - 2\,\underset{\cdot\cdot\cdot}{g}(t-1) - 3\,\underset{\cdot\cdot}{g}(t-1) + \underset{\cdot\cdot}{g}(t-2)$$

$$= \begin{cases} 0 & \text{for } t \leqslant 0 \\ 2\left(1 - e^{-t} - t + \dfrac{t^2}{2}\right) & \text{for } 0 \leqslant t \leqslant 1 \\ 2\left(1 - e^{-t} - t + \dfrac{t^2}{2}\right) - 2\left(1 - e^{-(t-1)} - (t-1) + \dfrac{(t-1)^2}{2}\right) & \\ \quad - 3(e^{-(t-1)} - 1 - (t-1)) & \text{for } 1 \leqslant t \leqslant 2 \\ -2e^{-t} + e^{-(t-1)} + 3 - 2t + (t-1) + t^2 - (t-1)^2 & \\ \quad + e^{-(t-2)} - 1 + (t-2) & \text{for } 2 \leqslant t \end{cases}$$

$$= \begin{cases} 0 & \text{for } t \leqslant 0 \\ 2\left(1 - e^{-t} - t + \dfrac{t^2}{2}\right) & \text{for } 0 \leqslant t \leqslant 1 \\ -e^{-t}(2 - e) + 3 - t & \text{for } 1 \leqslant t \leqslant 2 \\ e^{-t}(e^2 + e - 2) & \text{for } 2 \leqslant t \end{cases} \tag{4-46}$$

4-2 Block Diagram System Representation

The operations of addition and convolution are often represented by a diagram. There are several different closely related diagram forms appearing in the literature. The block diagram is used herein. The signal flow graph, which is just a generalization of the block diagram, and the Coates flow graph, which has certain computational advantages, are others (see Reference 20). In a block diagram, a line with an arrow represents a signal. When the signal enters a block with a time function written thereupon, the signal and the function of the block are convolved. The result of this convolution is a signal represented by a line with an arrow directed away from the block, Specifically,

$$y(t) = x * h(t)$$

is represented by Figure 4-1(a). When two or more signals come together

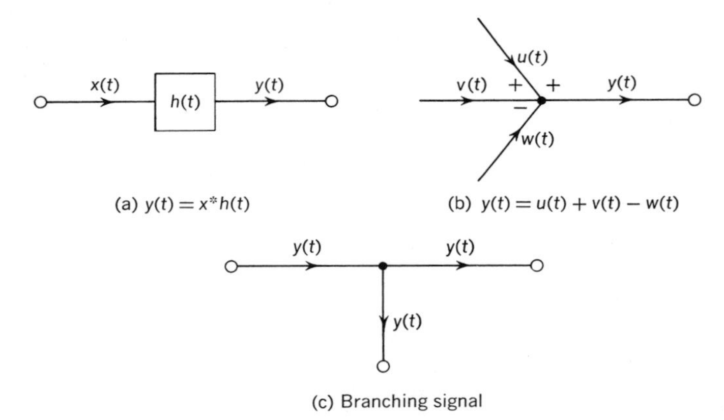

(a) $y(t) = x^* h(t)$

(b) $y(t) = u(t) + v(t) - w(t)$

(c) Branching signal

Figure 4-1 Block diagram symbols.

at a point, they are added or subtracted as per the indicated symbol, just as in Figure 2-7(c) above. The resultant is shown as an outgoing signal. Thus,

$$y(t) = u(t) + v(t) - w(t)$$

is represented by Figure 4-1(b). Finally, when a signal line branches, it means the same signal goes to several equations, as in Figure 4-1(c).

The most common block diagram is the single input, single output, single negative feedback loop diagram of Figure 4-2. For this diagram

$$x(t) = u(t) - y(t) \tag{4-47a}$$

$$y(t) = h * x(t) \tag{4-47b}$$

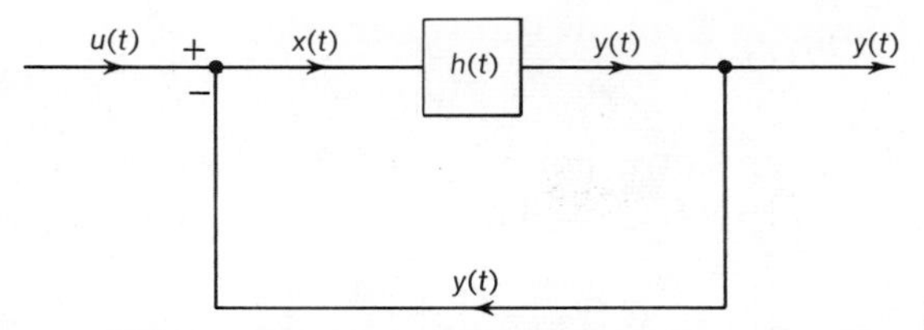

Figure 4-2 A simple feedback diagram.

Using Equation (4-47a) to eliminate x in (4-47b) gives

$$y(t) = h * u(t) - h * y(t) \tag{4-48}$$

Since convolution is distributive and commutative, we may put the terms involving y on the left and write

$$[\delta + h] * y(t) = h * u(t) \tag{4-49}$$

To solve this equation for y we need a way to find a function g such that

$$g * [\delta + h](t) = \delta(t) \tag{4-50}$$

Then we could convolve both sides of Equation (4-49) with g and get

$$y(t) = g(t) * h(t) * u(t) \tag{4-51}$$

If $h(t)$ is zero for $t < 0$, the system represented by the block is said to be *causal*.[4] In this case, $g(t)$ can be found by Laplace transforms in the usual sense. In this text our primary interest is in the special class of causal systems whose response functions were discussed in Chapter 3. In this case for $t \geqslant 0, h(t)$ is a sum of exponentials, polynomials times exponentials, delta functions and their derivatives. Formula (3-39) represents the convolution of v with

$$h(t) = \sum_{i=1}^{n} h_{ijk}\, e^{p_i t} + D_{jk}\, \delta(t) + E_{jk}\, \dot{\delta}(t)$$

Steady-state analysis. In the case of the systems of Chapter 3, a table of transforms is not needed to find $g(t)$ from a given $h(t)$. The symbolic steady-state methods of Section 3-3, which are equivalent to Laplace transforms, can be used. The final result is readily recognized from a partial fraction expansion. As an example for $t \geqslant 0$, let

$$h(t) = 11\, e^{-t} + 6te^{-t} + 6\delta(t) + \dot{\delta}(t) \tag{4-52}$$

The corresponding steady-state transfer function is

---

[4] There is a general definition of causality that applies to systems wherein the input-output relation is not a convolution. For convolutions the above is equivalent to this definition.

$$H(j\omega) = \frac{11}{j\omega + 1} + \frac{6}{(j\omega + 1)^2} + 6 + j\omega \tag{4-53}$$

The transfer function corresponding to $\delta(t) + h(t)$ for Equation (4-50) is found by adding 1 to $H(j\omega)$. The transfer function for $g(t)$ is just

$$G(j\omega) = \frac{1}{1 + H(j\omega)} \tag{4-54}$$

or

$$\begin{aligned} G(j\omega) &= \frac{(j\omega + 1)^2}{j\omega(j\omega + 1)^2 + 7(j\omega + 1)^2 + 11(j\omega + 1) + 6} \\ &= \frac{(j\omega + 1)^2}{(j\omega)^3 + 9(j\omega)^2 + 26(j\omega) + 24} \\ &= \frac{(j\omega + 1)^2}{(j\omega + 2)(j\omega + 3)(j\omega + 4)} \\ &= \frac{\frac{1}{2}}{j\omega + 2} - \frac{4}{j\omega + 3} + \frac{\frac{9}{2}}{j\omega + 4} \end{aligned} \tag{4-55}$$

Then for $t \geqslant 0$

$$g(t) = \tfrac{1}{2}\, e^{-2t} - 4\, e^{-3t} + \tfrac{9}{2}\, e^{-4t} \tag{4-56}$$

The block diagrams discussed above have one input signal and one output signal per block. The techniques apply equally well when there are several input and several output signals. In such cases we just consider an input signal vector and an output signal vector. The response function in the block is a matrix of functions similar to the single response functions $h(t)$ above.

For systems characterized by Equations (3-35), the number of inputs is the dimension of $\mathbf{v}$. The number of outputs is the dimension of $\mathbf{y}$. In a block diagram the box is described by a matrix $\underset{\sim}{h}(t)$; and the input output relation is

$$\mathbf{y}(t) = \underset{\sim}{h} * \mathbf{v}(t) \tag{4-57}$$

The rule for the matrix convolution operation in Equation (4-57) is the same as that for matrix multiplication, only convolution replaces multiplication in each operation; for example,

$$y_i(t) = \sum_{j=1}^{m} h_{ij} * v_j(t)$$

For the system in question the components of $\underset{\sim}{h}(t)$ are all causal. The formulas for $t \geqslant 0$ are sums of exponentials, polynomials times exponentials, delta functions to take care of $\underset{\sim}{D}$ in (3-35), and derivatives of delta functions to take care of $\underset{\sim}{E}$. Thus, most of the manipulations with block diagrams representing systems characterized by equations such as (3-35) can be carried out by manipulation of steady-state transfer functions and then partial fraction expansions. For such systems there is no need for formal Laplace transforms for manipulating block diagrams.

State variables for block diagram. Often one is interested in going from a block diagram to a set of state variable equations. If the block represents a system that can be characterized by normal form equations with constant coefficients, then $h(t)$ is causal and $H(j\omega)$ is a ratio of polynomials. To be specific, consider a single-input, single-output system characterized by a transfer function and its corresponding response function in the form

$$H(j\omega) = a(j\omega) + a_0 + \frac{a_1}{j\omega + p_1} + \frac{a_2}{j\omega + p_2} + \cdots + \frac{a_n}{(j\omega + p_n)}$$
$$h(t) = a\,\dot{\delta}(t) + a_0\,\delta(t) + a_1 e^{-p_1 t} + \cdots + a_n e^{-p_n t} \tag{4-58}$$

To keep from having to detail special cases, we assume the p_i are all distinct.

The simplest matrix with eigenvalues $-p_1, -p_2, \cdots, -p_n$ is a diagonal matrix with these numbers on the diagonal. Thus, a possible $\underset{\sim}{A}$ matrix for the system in question is

$$\underset{\sim}{A} = \begin{bmatrix} -p_1 & 0 & 0 & \cdots & 0 \\ 0 & -p_2 & 0 & \cdots & 0 \\ 0 & 0 & -p_3 & \cdots & 0 \\ \vdots & \vdots & \vdots & \vdots & \vdots \\ 0 & 0 & 0 & \cdots & -p_n \end{bmatrix} \tag{4-59}$$

Since the system has only one input and one output, the vectors **v** and **y** in Equation (3-35) have only one component—they are scalars. Thus, the matrix $\underset{\sim}{B}$ is an n-vector, the matrix $\underset{\sim}{C}$ is $1 \times n$—a transposed vector; and the matrixes $\underset{\sim}{D}$ and $\underset{\sim}{E}$ are 1×1—scalars. We now must find the constants in **B**, $\underset{\sim}{C}$, D, and E.

With the system with $\underset{\sim}{A}$ matrix (4-59), each state variable is characterized by the independent first-order equations

$$\dot{x}_i = p_i\, x_i + b_i\, v \tag{4-60}$$

The zero-state response for this equation is

$$x_i = \int_{t_0}^{t} b_i e^{-p_i(t-\tau)}\, v(\tau)\, d\tau \tag{4-61}$$

If p_i has a positive real part, then so long as $v(t)$ is bounded on $(-\infty, \infty)$, it is reasonable to consider values back to $t_0 \to -\infty$. Then Equation (4-61) becomes

$$x_i(t) = \int_{-\infty}^{t} b_i e^{-p_i(t-\tau)} \, v(\tau) \, d\tau \tag{4-62}$$

With Equation (4-58) for the transfer function and Equation (4-62) for the zero-state response of each state variable, the zero-state response for the system is

$$y(t) = \sum_{i=1}^{n} \int_{-\infty}^{t} c_i \, b_i e^{-p_i(t-\tau)} \, v(\tau) \, d\tau + D \, v(t) + E \, \dot{v}(t) \tag{4-63}$$

Comparing Equation (4-63) to the form of response of a block diagram with input $v(t)$ and response function given by (4-58), we see that so long as $c_i b_i = a_i$, $i = 1, 2, \cdots, n$ and $a_0 = D$, $a = E$, the forms (4-58) and (3-35) are equivalent so far as the zero-state response is concerned.

As an illustrative example, suppose for $t \geqslant 0$

$$h(t) = e^{-t} \cos t + \delta(t) \tag{4-64}$$

Then

$$H(j\omega) = \frac{\frac{1}{2}}{j\omega + 1 + j} + \frac{\frac{1}{2}}{j\omega + 1 - j} + 1 \tag{4-65}$$

The corresponding $\underset{\sim}{A}$ matrix is

$$\underset{\sim}{A} = \begin{bmatrix} -1-j & 0 \\ 0 & -1+j \end{bmatrix} \tag{4-66}$$

The **B** and **C** vectors can have any value provided that $b_1 \, c_1 = b_2 \, c_2 = \frac{1}{2}$. Let us choose $b_1 = 1$, $b_2 = \frac{1}{2}$. The scalar D is 1 and E is zero in (3-35). The final result is the state variable equation

$$\begin{aligned} \dot{\mathbf{x}} &= \begin{bmatrix} -1-j & 0 \\ 0 & -1+j \end{bmatrix} \mathbf{x} + \begin{bmatrix} 1 \\ \frac{1}{2} \end{bmatrix} v \\ y &= [\, \tfrac{1}{2} \;\; 1 \,] \, \mathbf{x} + v \end{aligned} \tag{4-67}$$

A second way to generate a set of state variables from a rational transfer function is to write the transfer function in the form

$$H(j\omega) = \frac{A_0 + A_1(j\omega) + \cdots + A_m(j\omega)^m}{B_0 + B_1(j\omega) + \cdots + B_n(j\omega)^n} \tag{4-68}$$

If Equations (4-68) and (4-58) are equivalent, then $m = n + 1$. By assuming an excitation $v(t) = e^{j\omega t}$, one can readily show that the transfer function (4-68) can come from the nth-order differential equation

$$B_n D^n y + B_{n-1} D^{n-1} y + \cdots + B_1 \dot{y} + B_0 y = A_m D^m v + \cdots + A_1 \dot{v} + A_0 v \tag{4-69}$$

Normal form equations equivalent to (4-69) can be constructed using the method discussed in Section 2-4.

For the block diagram with response function (4-64) and transfer function (4-65), the constants in Equation (4-68) are

$$H(j\omega) = \frac{(j\omega)^2 + (3j\omega) + 3}{(j\omega)^2 + 2j\omega + 2} \tag{4-70}$$

The corresponding differential equation in the form (4-69) is

$$\ddot{y} + 2\dot{y} + 2y = \ddot{v} + 3\dot{v} + 3v$$

Since (2-15) only applies when the highest derivative on the right is less than that on the left, we must modify the procedure for (4-70). We could write

$$H(j\omega) = 1 + \frac{j\omega + 1}{(j\omega)^2 + 2j\omega + 2}$$

Then $y(t)$ can be written as

$$y(t) = v(t) + y_1(t) \tag{4-71}$$

where $y_1(t)$ is a solution to the differential equation

$$\ddot{y}_1 + 2\dot{y}_1 + 2y_1 = \dot{v} + v \tag{4-72}$$

Now by Equations (2-14) and (2-15), (4-72) is equivalent to the normal form equations

$$\dot{x}_1 = x_2 + v$$

$$\dot{x}_2 = -2x_1 - 2x_2 - v$$

From (4-71) and the fact that y_1 is the state variable x_1, the final set of equations for the system is

$$\dot{\mathbf{x}} = \begin{bmatrix} 0 & 1 \\ -2 & -2 \end{bmatrix} \mathbf{x} + \begin{bmatrix} 1 \\ -1 \end{bmatrix} v$$
$$y = x_1 + v \tag{4-73}$$

Change of state variables. For a given response function or transfer function, infinitely many equivalent sets of normal form equations can be constructed. Since the state variables x_i do not appear in the input-output specification, any nonsingular linear transformation of a given set of x_i could be used to construct another set. That is, for a given set x_i, new variables w_i can be constructed by

$$\mathbf{w} = \underset{\sim}{K}\,\mathbf{x} \tag{4-74}$$

The determinant of $\underset{\sim}{K}$ must be nonzero so that the x_i can be found from the w_i by

$$\mathbf{x} = \underset{\sim}{K}^{-1}\,\mathbf{w} \tag{4-75}$$

Now from a given normal form equivalent with state variables x_i, a second equivalent with variables w_i can be obtained by substituting Equation (4-75) for $\mathbf{x}$. That is, given

$$\begin{aligned} \dot{\mathbf{x}} &= \underset{\sim}{A}\,\mathbf{x} + \underset{\sim}{B}\,\mathbf{v} \\ \mathbf{y} &= \underset{\sim}{C}\,\mathbf{x} + \underset{\sim}{D}\,\mathbf{v} + \underset{\sim}{E}\,\dot{\mathbf{v}} \end{aligned} \tag{4-76}$$

we first write

$$\begin{aligned} \underset{\sim}{K}^{-1}\,\dot{\mathbf{w}} &= \underset{\sim}{A}\,\underset{\sim}{K}^{-1}\,\mathbf{w} + \underset{\sim}{B}\,\mathbf{v} \\ \mathbf{y} &= \underset{\sim}{C}\,\underset{\sim}{K}^{-1}\,\mathbf{w} + \underset{\sim}{D}\,\mathbf{v} + \underset{\sim}{E}\dot{\mathbf{v}} \end{aligned} \tag{4-77}$$

Then

$$\begin{aligned} \dot{\mathbf{w}} &= \underset{\sim}{K}\,\underset{\sim}{A}\,\underset{\sim}{K}^{-1}\,\mathbf{w} + \underset{\sim}{K}\,\underset{\sim}{B}\,\mathbf{v} \\ \mathbf{y} &= \underset{\sim}{C}\,\underset{\sim}{K}^{-1}\,\mathbf{w} + \underset{\sim}{D}\,\mathbf{v} + \underset{\sim}{E}\,\dot{\mathbf{v}} \end{aligned} \tag{4-78}$$

This is also an equivalent set of normal form equations so far as input and output are concerned. The matrix $\underset{\sim}{K}\,\underset{\sim}{A}\,\underset{\sim}{K}^{-1}$ has the same eigenvalues as those of $\underset{\sim}{A}$.

As an illustration of the change of state variables for a given response function, let us find the $\underset{\sim}{K}$ that transforms the $\underset{\sim}{A}$ matrix of Equation (4-73) into that of Equation (4-66). The equation for $\underset{\sim}{K}$ is

$$\underset{\sim}{K}\begin{bmatrix} 0 & 1 \\ -2 & -2 \end{bmatrix}\underset{\sim}{K}^{-1} = \begin{bmatrix} -1-j & 0 \\ 0 & -1+j \end{bmatrix} \tag{4-79}$$

Postmultiplying both sides of Equation (4-75) by $\underset{\sim}{K}$ gives

$$\begin{bmatrix} -2\,k_{12} & k_{11} - 2\,k_{12} \\ -2\,k_{22} & k_{21} - 2\,k_{22} \end{bmatrix} = -\begin{bmatrix} (1+j)\,k_{11} & (1+j)\,k_{12} \\ (1-j)\,k_{21} & (1-j)\,k_{22} \end{bmatrix} \tag{4-80}$$

Equating corresponding elements shows that

$$\begin{aligned} k_{12} &= \frac{1+j}{2}\,k_{11} \\ k_{21} &= \frac{2}{1-j}\,k_{22} \end{aligned} \tag{4-81}$$

Any values for k_{11} and k_{22} will satisfy. This freedom is to be expected since for a particular $\underset{\sim}{A}$ matrix of (4-73) there is freedom in choosing the $\underset{\sim}{B}$ and $\underset{\sim}{C}$ matrixes. For the present let us choose $k_{11} = k_{22} = 1$. Then

$$\underset{\sim}{K} = \begin{bmatrix} 1 & \dfrac{1+j}{2} \\ \dfrac{2}{1-j} & 1 \end{bmatrix} \tag{4-82}$$

The term $\underset{\sim}{K}\ \underset{\sim}{B}$ in (4-78) gets its $\underset{\sim}{B}$ matrix from (4-73) for this example. Specifically,

$$\underset{\sim}{K}\ \underset{\sim}{B} = \begin{bmatrix} 1 & \dfrac{1+j}{2} \\ \dfrac{2}{1-j} & 1 \end{bmatrix} \begin{bmatrix} 1 \\ -1 \end{bmatrix} = \begin{bmatrix} \dfrac{1-j}{2} \\ \dfrac{1+j}{1-j} \end{bmatrix} \tag{4-83}$$

This is not the $\underset{\sim}{B}$ matrix of (4-67). Let us proceed with the computation of $\underset{\sim}{C}\ \underset{\sim}{K}^{-1}$ for (4-78) and show that the result is the $\underset{\sim}{C}$ matrix we would have obtained for (4-67) when (4-83) is the $\underset{\sim}{B}$ matrix.

With $\underset{\sim}{K}$ from (4-82),

$$\underset{\sim}{K}^{-1} = \begin{bmatrix} \dfrac{1+j}{2} & \dfrac{j}{2} \\ j & \dfrac{1+j}{2} \end{bmatrix} \tag{4-84}$$

Using the $\underset{\sim}{C}$ matrix from (4-73) in (4-78) we get

$$C\ K^{-1} = [1 \quad 0] \begin{bmatrix} \dfrac{1+j}{2} & \dfrac{j}{2} \\ j & \dfrac{1+j}{2} \end{bmatrix} = \begin{bmatrix} \dfrac{1+j}{2} & \dfrac{j}{2} \end{bmatrix} \tag{4-85}$$

As expected, the product of the first term of the right side of (4-83) with the first term of that of (4-85) is $\frac{1}{2}$. The same is true of the respective second terms. Thus the transformation $\underset{\sim}{K}$ applied to the state equations (4-73) gives the equations

$$\begin{aligned} \dot{\mathbf{x}} &= \begin{bmatrix} -1-j & 0 \\ 0 & -1+j \end{bmatrix} \mathbf{x} + \begin{bmatrix} \dfrac{1-j}{2} \\ \dfrac{1+j}{1-j} \end{bmatrix} v \\ y &= \begin{bmatrix} \dfrac{1+j}{2} & \dfrac{j}{2} \end{bmatrix} \mathbf{x} + v \end{aligned} \tag{4-86}$$

Equations (4-86) define the same system as (4-73) and (4-67).

As pointed out in the beginning of this section a block diagram interconnection of rational transfer functions can be solved to give a single system response function by algebraic manipulation of the transfer functions and then a partial fraction expansion. A second approach is to describe each block by a set of state variables. The algebraic operations at the summing junctions show how the normal form equations plus their associated input-output equations are combined. The result is a single set of normal form equations plus input output equations for the entire system. The demonstration of this technique is left to the problems.

4-3 The Total Response—Zero State Plus Zero Input

Thus far in this chapter we have discussed systems whose input-output relation is a convolution with no reference to the possibility of a zero-input response. Thus, we have been discussing only the zero-state response if these systems have a total response that can be separated into a zero-state term plus a zero-input term. In the definition of *state at* t_0 in Chapter 1, the input was known only between t_0 and t_1. Let us consider some relations between systems described by convolutions and the definition of state, linearity, and time-invariance from Chapter 1.

Linearity, time-invariance and the state at t_0. First let us consider a system with one input and one output with total response given by the convolution

$$y(t) = \int_{-\infty}^{\infty} h(t-\tau)\, v(\tau)\, d\tau \tag{4-87}$$

Separating out that part of the response at t_1 that depends on the input between t_0 and t_1 gives

$$y(t_1) = \int_{-\infty}^{t_0} h(t_1-\tau)\, v(\tau)\, d\tau + \int_{t_1}^{\infty} h(t_1-\tau)\, v(\tau)\, d\tau + \int_{t_0}^{t_1} h(t_1-\tau)\, v(\tau)\, d\tau \tag{4-88}$$

If the state at t_0 is to have meaning, it must be derivable from the first two integrals of Equation (4-88).

If the system is not causal, then the state at t_0 depends on the inputs prior to t_0 and those after t_1. When the system is causal, then only prior inputs are needed to determine the state at t_0. Even for a causal system characterized by a convolution, the entire input from $-\infty$ to t_0 must be known to determine the total output for any t_1. The zero-input term depends on both t_0 and t_1 in a way that involves the input differently for different t_1. In one special case, a case that includes the lumped systems of Chapters 2 and 3, the state at t_0 does not require knowledge of the input for all $t < t_0$. That case occurs when the response function is separable into a sum of terms of the form of the product of a function of t with a function of τ. That is, when

$$h(t-\tau) = \sum_{i=1}^{n} f_i(t)\; g_i(\tau) \tag{4-89}$$

In this case Equation (4-88) becomes

$$y(t_1) = \sum_{i=1}^{n} f_i(t_1) \int_{-\infty}^{t_0} g_i(\tau)\; v(\tau)\; d\tau + \int_{t_0}^{t_1} h(t_1-\tau)\; v(\tau)\; d\tau \tag{4-90}$$

Since $f_i(t)$ depends on the system and not on the input, the state at t_0 is now the set of n numbers $\int_{-\infty}^{t_0} g_i(\tau)\; v(\tau)\; d\tau$.

Any system characterized by Equation (4-87) is both linear and time-invariant. For linearity let us apply the definition with respect to the input signal on the interval (t_0, t_1). The first computation is the determination of $S_{oo}(t)$, the output when the input is zero. It is

$$S_{oo}(t) = \int_{-\infty}^{t_0} h(t-\tau)\; v(\tau)\; d\tau + \int_{t_1}^{\infty} h(t-\tau)\; v(\tau)\; d\tau \tag{4-91}$$

For an input signal $S_{i1}(t)$ on the interval (t_0, t_1), the total output is

$$S_{o1}(t) = S_{oo}(t) + \int_{t_0}^{t_1} S_{i1}(\tau)\; h(t-\tau)\; d\tau \tag{4-92}$$

Similarly, for input signal $S_{i2}(t)$, the total output is

$$S_{o2}(t) = S_{oo}(t) + \int_{t_0}^{t_1} S_{i2}(\tau)\; h(t-\tau)\; d\tau \tag{4-93}$$

When the input on the interval (t_0, t_1) is $A_1\, S_{i1}(t) + A_2\, S_{i2}(t)$, the output is

$$S_o(t) = S_{oo}(t) + \int_{t_0}^{t_1} [A_1\, S_{i1}(\tau) + A_2\, S_{i2}(\tau)]\; h(t-\tau)\; d\tau \tag{4-94}$$

A comparison of Equation (4-94) with (4-92) and (4-93) shows that the system is linear since

$$S_o(t) = A_1[S_{o1}(t) - S_{oo}(t)] + A_2[S_{o2}(t) - S_{oo}(t)] + S_{oo}(t) \tag{4-95}$$

To show time-invariance for the system characterized by (4-87) we must compare the response with input $v(t)$ to that with input $v(t+a)$. If we use formula (4-88) as the separation of zero-input and zero-state response, there are problems in applying the definition. These arise because substituting $v(t+a)$ in the formula changes the zero-input term and consequently changes the state at t_0. In the definition it was assumed that the state at t_0 was the same for both the input $v(t)$ and the input $v(t+a)$. The only situation to which the definition can be applied by a simple substitution is the case where t_0 is taken as $-\infty$ and the zero-input response is zero. Then (4-87) represents only the zero-state response. In this case the output for input $v(t+a)$ is

$$x(t) = \int_{-\infty}^{\infty} h(t-\tau)\, v(\tau + a)\, d\tau \tag{4-96}$$

By the change of variable $\lambda = \tau + a$ in the integral

$$x(t) = \int_{-\infty}^{\infty} h(t-\lambda+a)\, v(\lambda)\, d\lambda = y(t+a) \tag{4-97}$$

where $y(t)$ is the output given by Equation (4-87). Thus, the system is time-invariant.

Thus far in this section we have considered only the situation where the input-output relation gave the total response when the input was known for all time. Although only the single-input–single-output case was discussed explicitly, the statements all generalize to a multiple input and output system characterized by a matrix of response functions such as the system of Equation (4-57). In general, the response function matrix that can be ascertained from input-output measurements on a system may not be sufficient to ascertain the total response. For lumped linear systems whose input-output relations are of the form (3-39), such a situation occurs when one of the h_{ijk} is zero. Then one of the natural frequencies does not appear in the forced response, but it may still appear in the natural response.

Controllability and observability. The problem of getting a complete description from input-output measurements is tied up with two questions as follows:

1. If a system is in an arbitrary state at t_0, can a set of inputs be found so that if these inputs are applied between t_0 and t_1 the system will be in the zero state at t_1?
2. If the output of a system is observed on the interval (t_0, t_1), can the state of the system at t_0 be ascertained?

When the answer to the first question is yes, then the system is said to be *controllable*. If the answer to the second question is yes the system is said to be *observable*.

A general discussion of controllability and observability for linear systems characterized by convolutions is beyond the scope of this text. For the special case of lumped, linear, time-invariant systems with distinct natural frequencies, the two questions are fairly easy to answer. To be specific, let us consider the one-input—one-output system characterized by the response function and transfer function (4-58). As pointed out above, one possible set of normal form equations for this system is

$$\begin{aligned} \dot{\mathbf{x}} &= \underset{\sim}{A}\ \mathbf{x} + \mathbf{B}\ v \\ y &= \underset{\sim}{C}\ \mathbf{x} + D\ v + E\ \dot{v} \end{aligned} \tag{4-98}$$

where $\underset{\sim}{A}$ is an $n \times n$ diagonal matrix with diagonal elements $-p_i$; $\mathbf{B}$ is a vector with components b_i; $\underset{\sim}{C}$ is a row matrix with components c_i such that $b_i c_i = a_i$ in (4-58). Let us next assume that Equations (4-98) completely characterize the system. The total response is then

$$y(t) = \int_{t_0}^{t} h(t-\tau)\ v(\tau)\ d\tau + \sum_{i=1}^{n} c_i x_i(t_0)\ e^{-p_i(t-t_0)} \tag{4-99}$$

To show that the system is controllable let us assume $\mathbf{x}(t_0)$ given and show that for any $t_1 \neq t_0$ we can find an input $v(t)$ on the interval (t_0, t_1) so that $\mathbf{x}(t_1) = \mathbf{0}$. With the special form of system equations (4-98) each state variable is given by

$$x_i(t) = \int_{t_0}^{t} b_i e^{-p_i(t-\tau)}\ v(\tau)\ d\tau + x_i(t_0)\ e^{-p_i(t-t_0)} \tag{4-100}$$

The problem is to find a function v defined on (t_0, t_1) such that each $x_i(t_1)$ as given by (4-100) is zero.

There is no unique v that sets all n of the $x_i(t_1)$ to zero. Since these are n constraints, almost any v with n independently set conditions will do. One possibility is the piecewise constant function

$$v(t-t_0) = K_k \text{ for } (k-1)\left(\frac{t_1-t_0}{n}\right) \leqslant t - t_0 \leqslant k\left(\frac{t_1-t_0}{n}\right) \qquad k = 1, 2, \cdots, n \tag{4-101}$$

The problem now is the determination of the K_k so that $\mathbf{x}(t_1) = 0$. With $v(t)$ given by (4-101), each $x_i(t_1)$ is given by

$$x_i(t_1) = e^{-p_i t_1} \left[\frac{b_i}{p_i} \sum_{k=1}^{n} K_k\left(e^{\frac{p_i k(t_1-t_0)}{n}} - e^{\frac{p_i(k-1)(t_1-t_0)}{n}}\right) + x_i(t_0)\ e^{-p_i t_0}\right] \tag{4-102}$$

Since each $x_i(t_1)$ must be zero, the K_j's are determined by the algebraic equations

$$\underset{\sim}{M}\ \mathbf{K} = \mathbf{L} \tag{4-103}$$

where $\underset{\sim}{M}$ is an $n \times n$ matrix with elements

$$m_{ik} = \frac{b_i}{p_i}\left[e^{\frac{p_i k(t_1 - t_0)}{n}} - e^{\frac{p_i(k-1)(t_1 - t_0)}{n}} \right]$$

K is an n-vector with components K_i; **L** is an n-vector with components $L_i = x_i(t_0)\, e^{-p_i t_0}$.

Equations (4-103) can be solved for the K_i's by any convenient method, provided $\underset{\sim}{M}$ is nonsingular. Since the p_i's are assumed distinct, the rows of $\underset{\sim}{M}$ are certainly linearly independent. There is possibility that in a particular case the numbers are such that the columns are not independent. However, since the intervals over which $v(t)$ takes on the various constant values can be specified independently, a change in the intervals in (4-101) can be made if $\underset{\sim}{M}$ should be singular for uniform intervals.

For the system to be observable the state $\mathbf{x}(t_0)$ must be obtainable from output measurements. Since all the $x_i(t_0)$ appear in the output (4-99), n-independent measurements are sufficient. If there is a known input during the period of observation, the forced response term can be computed and subtracted from the measurements. Then the n values of the zero-input term are readily used to solve for the $x_i(t_0)$.

A system is not both controllable and observable if the number of natural frequencies in the response function is less than the dimension of the state vector. For this situation we again consider a system characterized by Equations (4-98) and (4-58) but now the $\underset{\sim}{A}$ matrix in (4-98) is $q \times q$ and diagonal with $q = n + m + r$. Furthermore, **B** has components $b_i = 0$ for $n + 1 \leq i \leq n + m$, and $\underset{\sim}{C}$ has components $c_i = 0$ for $(n + m + 1) \leq i \leq q$. All other b_i and c_i are nonzero and $b_i c_i = a_i$ in (4-58) for $i = 1, 2, \cdots, n$. The response function is the same as before. The total response is formally as in (4-99), but because of the zeros in $\underset{\sim}{C}$, the vector $\mathbf{x}(t_0)$ cannot be determined from q measurements on y. For those values of i for which $b_i = 0$, the corresponding $x_i(t)$ is not related to $v(t)$ at all. Therefore, these $x_i(t)$ cannot be driven to zero by any choice of $v(t)$. This system is neither controllable nor is it observable.

In general, a lumped, linear, time-invariant system is not both controllable and observable if the natural frequencies do not appear as poles in the transfer function matrix. In order to investigate the controllability and observability of a system, one must have a complete description of the system—a description that can be converted to a set of normal form equations with an output statement in the form

$$\begin{aligned} \dot{\mathbf{x}} &= \underset{\sim}{A}\,\mathbf{x} + \underset{\sim}{B}\,\mathbf{v} \\ \mathbf{y} &= \underset{\sim}{C}\,\mathbf{x} + \underset{\sim}{D}\,\mathbf{v} + \underset{\sim}{E}\,\dot{\mathbf{v}} \end{aligned} \qquad \textbf{(4-103)}$$

Only $\underset{\sim}{A}$, $\underset{\sim}{B}$, and $\underset{\sim}{C}$ need be investigated. In all cases the controllability and observability of the system can be ascertained by first finding a transformation that diagonalizes the $\underset{\sim}{A}$ matrix[5] as discussed in Section 4-3 above. From the diagonal form the required properties are readily investigated as was done with the single-input–single-output system (4-98).

Some of the problems associated with testing for controllability and observability are demonstrated by an examination of the simple 2-port circuit of Figure 4-3. The properties for this circuit depend on which

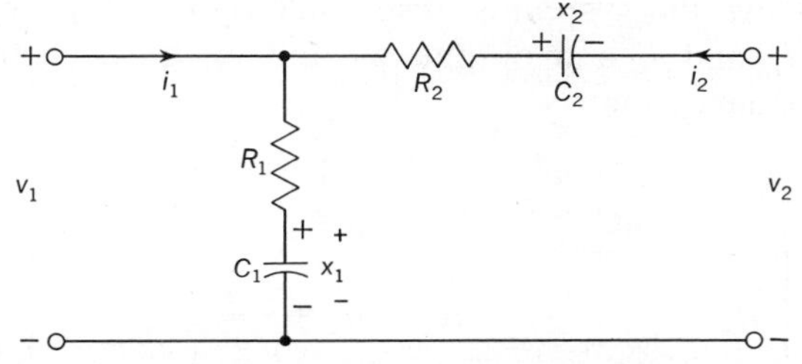

Figure 4-3 A circuit that demonstrates controllability and observability.

of the 4-port variables are inputs and which are outputs. Let us consider the circuit on an admittance basis, an impedance basis, and a hybrid basis.

On an admittance basis the transfer function matrix is the admittance matrix

$$\underset{\sim}{Y} = \begin{bmatrix} \dfrac{j\omega/R_1}{j\omega + \dfrac{1}{R_1C_1}} + \dfrac{j\omega/R_2}{j\omega + \dfrac{1}{R_2C_2}} & -\dfrac{j\omega/R_2}{j\omega + \dfrac{1}{R_2C_2}} \\ -\dfrac{j\omega/R_2}{j\omega + \dfrac{1}{R_2C_2}} & \dfrac{j\omega/R_2}{j\omega + \dfrac{1}{R_2C_2}} \end{bmatrix} \tag{4-104}$$

From this matrix we see that only one or the four transfer functions contains both natural frequencies. Thus, the system does not appear to be both controllable and observable at port-2. To see wherein the system fails, we must examine the progress of input signal to states to output signal.

From the circuit diagram a state variable description in the form (4-103) can be written directly. It is

$$\begin{aligned} \dot{x}_1 &= -\frac{1}{C_1R_1}x_1 + \frac{1}{C_1R_1}v_1 \\ \dot{x}_2 &= -\frac{1}{C_2R_2}x_2 + \frac{1}{C_2R_2}v_1 - \frac{1}{C_2R_2}v_2 \end{aligned} \tag{4-105}$$

---

[5] In the special case where the $\underset{\sim}{A}$ matrix cannot be diagonalized one must be content to work with the Jordan form. The basic procedure is the same. Reference 28, Chapter 11 has some additional tests for controllability and observability that are derivable from the concepts contained herein.

$$i_1 = -\frac{1}{R_1} x_1 - \frac{1}{R_2} x_2 + \left(\frac{1}{R_1} + \frac{1}{R_2}\right) v_1 - \frac{1}{R_2} v_2$$

$$i_2 = -\frac{1}{R_2} x_2 + \frac{1}{R_2} v_1 - \frac{1}{R_2} v_2$$

Clearly, the input v_1 can control both states, whereas v_2 can control only state 2. Also, the output i_1 depends on both states and i_2 depends on state two only. Thus, the system is observable by i_1 but not by i_2.

With inputs i_1 and i_2 and outputs v_1 and v_2, the transfer function matrix is the impedance matrix

$$\underset{\sim}{Z} = \begin{bmatrix} R_1 + \dfrac{1}{j\omega C_1} & R_1 + \dfrac{1}{j\omega C_1} \\ R_1 + \dfrac{1}{j\omega C_1} & R_1 + R_2 + \dfrac{1}{j\omega C_1} + \dfrac{1}{j\omega C_2} \end{bmatrix} \tag{4-106}$$

The system has two state variables, but the natural frequencies are both at zero. Thus, we can say nothing about controllability and observability from this impedance matrix.

The state variable description is

$$\begin{aligned} \dot{x}_1 &= \frac{1}{C_1} i_1 + \frac{1}{C_1} i_2 \\ \dot{x}_2 &= -\frac{1}{C_2} i_2 \\ v_1 &= x_1 + R_1 i_1 + R_1 i_2 \\ v_2 &= x_1 - x_2 + R_1 i_1 + (R_1 + R_2) i_2 \end{aligned} \tag{4-107}$$

In this case the $\underset{\sim}{A}$ matrix is the zero matrix. The same situation would arise for any diagonal matrix with repeated elements on the diagonal. At first glance it appears that input i_2 controls both states. With a little more thought, we are reminded that the demonstration of controllability above required distinct natural frequencies. From port-2, with $i_1 = 0$, we cannot put independent charges on C_1 and C_2. In fact, as a 1-port, the system has only one state variable, not two. Thus, the circuit is neither controllable or observable from port-2 alone. Using both inputs together we can control both states by first driving C_2 with i_2 and then setting C_1 with i_1. We can also observe both states if we use measurements from both ports.

On a hybrid parameter basis with i_1 and v_2 as inputs, the formulas are more complex. This case illustrates the problem of controllability and observability for systems when the $\underset{\sim}{A}$ matrix is not diagonal. Because of the complexity of the formulas with literal R's and C's, the notation is simplified with no loss of pedagogical value if specific numbers are used. To this end we choose $R_1 = R_2 = C_1 = 1, C_2 = 2$. Then the hybrid transfer function matrix is

$$\underset{\sim}{H}(j\omega) = \begin{bmatrix} \dfrac{(j\omega+1)(2j\omega+1)}{(4j\omega+3)j\omega} & \dfrac{j\omega+1}{4j\omega+3} \\ -\dfrac{2(j\omega+1)}{4j\omega+3} & \dfrac{j\omega}{4j\omega+3} \end{bmatrix} \tag{4-108}$$

Since the pole at zero appears only in H_{11}, the circuit is certainly not both controllable and observable from port-2 on this basis.

An investigation of controllability and observability requires the complete state variable description. It is

$$\begin{aligned} \dot{x}_1 &= -\tfrac{1}{2}x_1 + \tfrac{1}{2}x_2 + \tfrac{1}{2}i_1 + \tfrac{1}{2}v_2 \\ \dot{x}_2 &= \tfrac{1}{4}x_1 - \tfrac{1}{4}x_2 + \tfrac{1}{4}i_1 - \tfrac{1}{4}v_2 \\ v_1 &= \tfrac{1}{2}x_1 + \tfrac{1}{2}x_2 + \tfrac{1}{2}i_1 + \tfrac{1}{2}v_2 \\ i_2 &= -\tfrac{1}{2}x_1 + \tfrac{1}{2}x_2 - \tfrac{1}{2}i_1 + \tfrac{1}{2}v_2 \end{aligned} \tag{4-109}$$

From Equations (4-109) we can proceed either by diagonalizing the $\underset{\sim}{A}$ matrix and applying the formulas above or by getting a complete solution to the equations and applying the definitions of controllability and observability. We take the latter approach.

The state transition matrix for Equations (4-109) is

$$\underset{\sim}{\phi}(t, t_0) = \begin{bmatrix} \frac{1}{3} + \frac{2}{3}e^{-3/4(t-t_0)} & \frac{2}{3} - \frac{2}{3}e^{-3/4(t-t_0)} \\ \frac{1}{3} - \frac{1}{3}e^{-3/4(t-t_0)} & \frac{2}{3} + \frac{1}{3}e^{-3/4(t-t_0)} \end{bmatrix} \tag{4-110}$$

$$\begin{aligned} x_1(t) &= \int_{t_0}^{t} \tfrac{1}{3}i_1(\tau)\,d\tau + \int_{t_0}^{t} \tfrac{1}{6}i_1(\tau)\,e^{-3/4(t-\tau)}\,d\tau \\ &\quad + \int_{t_0}^{t} \tfrac{1}{2}v_2(\tau)\,e^{-3/4(t-\tau)}\,d\tau + \tfrac{1}{3}x_1(t_0) + \tfrac{2}{3}x_2(t_0) \\ &\quad + \tfrac{2}{3}x_1(t_0)\,e^{-3/4(t-t_0)} - \tfrac{2}{3}x_2(t_0)\,e^{-3/4(t-t_0)} \\ x_2(t) &= \int_{t_0}^{t} \tfrac{1}{3}i_1(\tau)\,d\tau - \int_{t_0}^{t} \tfrac{1}{12}i_1(\tau)\,e^{-3/4(t-\tau)}\,d\tau \\ &\quad - \int_{t_0}^{t} \tfrac{1}{4}v_2(\tau)\,e^{-3/4(t-\tau)}\,d\tau + \tfrac{1}{3}x_1(t_0) + \tfrac{2}{3}x_2(t_0) \\ &\quad - \tfrac{1}{3}x_1(t_0)\,e^{-3/4(t-t_0)} + \tfrac{1}{3}x_2(t_0)\,e^{-3/4(t-t_0)} \end{aligned} \tag{4-111}$$

From these two equations we see that i_1 can be choosen so that both x_1 and x_2 are controlled but that v_2 cannot drive the states so as to cancel the constant term $\frac{1}{3}\,x_1(t_0) + \frac{2}{3}\,x_2(t_0)$ that appears in the expression for each state.

For observability we look at the effect of the states on the outputs when there is no input. With $x_1(t_0)$ and $x_2(t_0)$ set, the zero-input response is

$$
\begin{aligned}
v_1(t) &= \tfrac{1}{2}\left[\tfrac{2}{3}x_1(t_0) + \tfrac{4}{3}x_2(t_0) + \tfrac{1}{3}\left(x_1(t_0) - x_2(t_0)\right) e^{-3/4(t-t_0)}\right] \\
i_2(t) &= \tfrac{1}{2}\left[x_2(t_0) - x_1(t_0)\right] e^{-3/4(t-t_0)}
\end{aligned}
\qquad \textbf{(4-112)}
$$

Clearly $x_1(t_0)$ and $x_2(t_0)$ can be obtained from two measurements on $v_1(t)$ but they cannot form measurements on $i_2(t)$. A knowledge of $i_2(t)$ gives a knowledge of only $(x_2(t_0) - x_1(t_0))$ but not the two states separately.

The ideas of controllability and observability apply to time-variant as well as time-invariant systems. For the latter case it takes more sophisticated mathematics to discuss the relation between normal form equations and response functions.

The state at t_0 as an additional input. As a final point we should note that the zero-input response can be displayed on either the circuit or analog computer diagrams by adding additional inputs with delta function or step function signal sources. The second-order systems of Figures

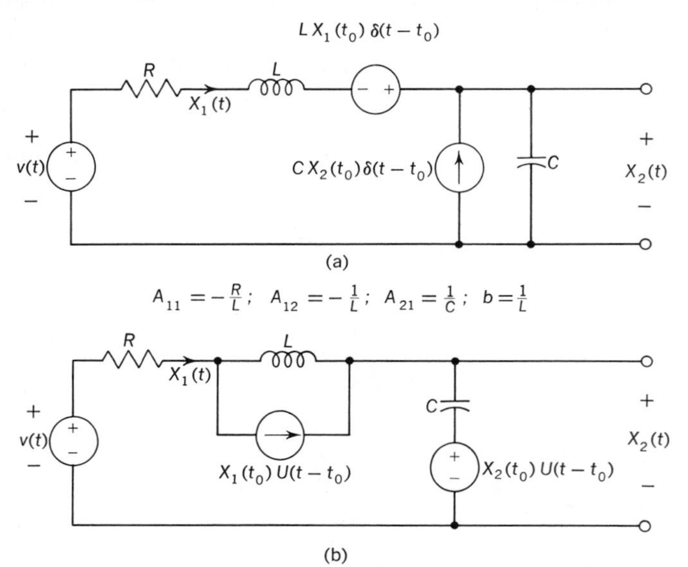

Figure 4-4 Circuit diagrams characterized by Equation (4-104). **(a)** Delta function sources for initial conditions. **(b)** Step function sources for initial conditions.

4-4 and 4-5 illustrate the procedure. For $t > t_0$ each diagram is characterized by

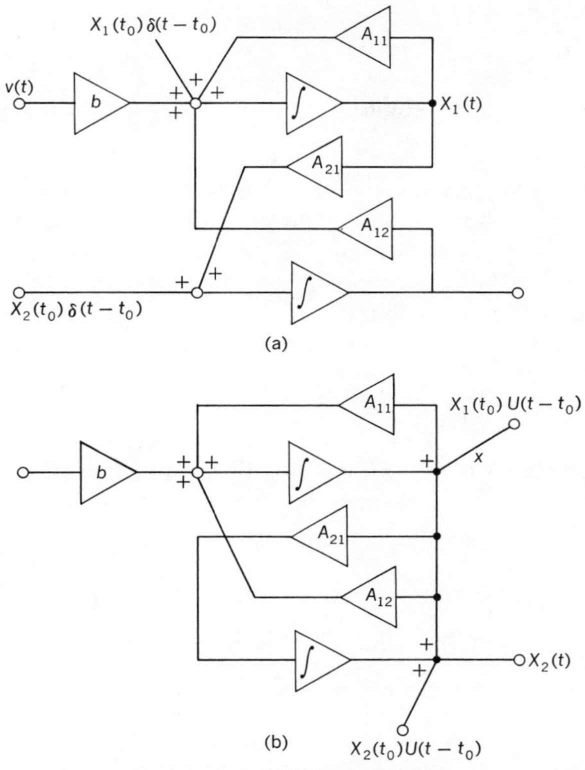

Figure 4-5 Analog computer diagram for Equation (4-104). (**a**) Delta function sources for initial conditions. (**b**) Step function sources for initial conditions.

$$\dot{x}_1 = a_{11}\, x_1 + a_{12}\, x_2 + bv$$

$$\dot{x}_2 = a_{21}\, x_1 \tag{4-113}$$

$$x_2 = \int_{t_0}^{t} [k_1\, e^{p_1(t-\tau)} + k_2\, e^{p_2(t-\tau)}]\; v(\tau)\; d\tau + K_1\, e^{p_1(t-t_0)} + K_2\, e^{p_2(t-t_0)}$$

where

$$k_1 = \frac{a_{21}b}{\sqrt{a_{11}{}^2 + 4a_{12}\, a_{21}}}$$

$$p_1 = \frac{a_{11}}{2} + \sqrt{\left(\frac{a_{11}}{2}\right)^2 + a_{12}\, a_{21}}$$

$$k_2 = \frac{-a_{21}b}{\sqrt{a_{11}{}^2 + 4a_{12}\, a_{21}}}$$

$$p_2 = \frac{a_{11}}{2} - \sqrt{\left(\frac{a_{11}}{2}\right)^2 + a_{12}\, a_{21}}$$

$$K_1 = \frac{-a_{12}}{\sqrt{a_{11}^2 + 4a_{12}\,a_{21}}}x_1(0) + \frac{a_{11} - \sqrt{a_{11}^2 + 4a_{12}\,a_{21}}}{2\sqrt{a_{11}^2 + 4a_{12}\,a_{21}}}x_2(0)$$

$$K_2 = \frac{-a_{12}}{\sqrt{a_{11}^2 + 4a_{12}\,a_{21}}}x_1(0) + \frac{a_{11} + \sqrt{a_{11}^2 + 4a_{12}\,a_{21}}}{2\sqrt{a_{11}^2 + 4a_{12}\,a_{21}}}x_2(0)$$

This idea of considering initial conditions as additional inputs is an easy way to remember the simplified procedure for computing the state transition matrix given in Section 3-3. In the frequency domain, a time-domain delta function becomes a constant. Thus, in the frequency domain, an equation with initial conditions takes the form of Equation (3-54) where the **U** is the state at zero.

PROBLEMS

4-1 Consider the system whose equations are given in Problem 3-4. Find the convolution expression for the zero-state response by using steady-state analysis techniques.

4-2 A single-input–single-output system has a response function

$$h(t) = Ke^{-(a+jb)t} + K^{*}e^{-(a-jb)t}$$

Find the output when the input is

$$f(t) = \begin{cases} 0 & \text{for } t < 0 \\ t^2 & \text{for } 0 \leq t < 1 \\ 2 - t & \text{for } 1 \leq t < 2 \\ 0 & \text{for } t > 2 \end{cases}$$

4-3 Let the input to a first-order system be the continuously differentiable pulse made of parabolic segments, as shown in Figure P4-3.

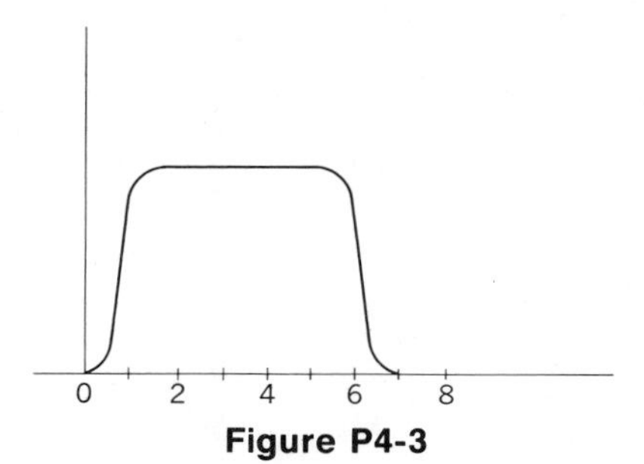

Figure P4-3

The pulse takes 2 μs to rise, is perfectly flat for 3 μs, and then falls symmetrically. Compute the output when the system time constant is (a) 5 μs; (b) 10 μs; (c) 50 μs. Sketch the results.

4-4 Consider the transfer function

$$H(j\omega) = \frac{2(j\omega) + 3}{(j\omega)^2 + 3j\omega + 2}$$

and its reciprocal

$$G(j\omega) = \frac{1}{H(j\omega)}$$

Compute the corresponding response functions $h(t)$ and $g(t)$. Recall $g(t)$ will contain the derivative of a delta function. For any continuous, integrable function $f(t)$ show $f * g * h(t) = f(t)$.

4-5 For the system represented by the block diagram of Figure P4-5,

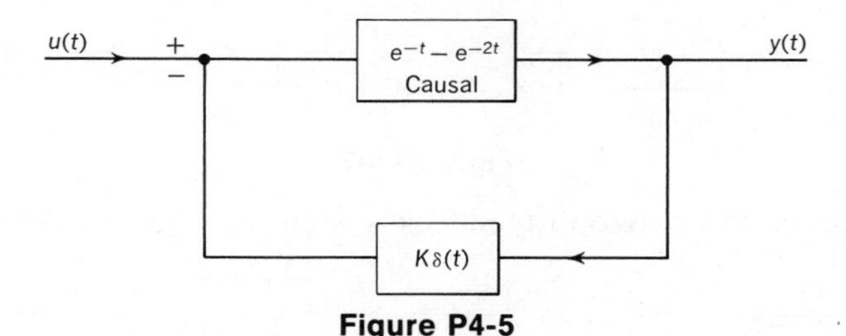

Figure P4-5

find the $\underset{\sim}{A}$, $\underset{\sim}{B}$, $\underset{\sim}{C}$, and $\underset{\sim}{D}$ matrixes for the over-all system and then select the value of K so that one of the natural frequencies is at zero.

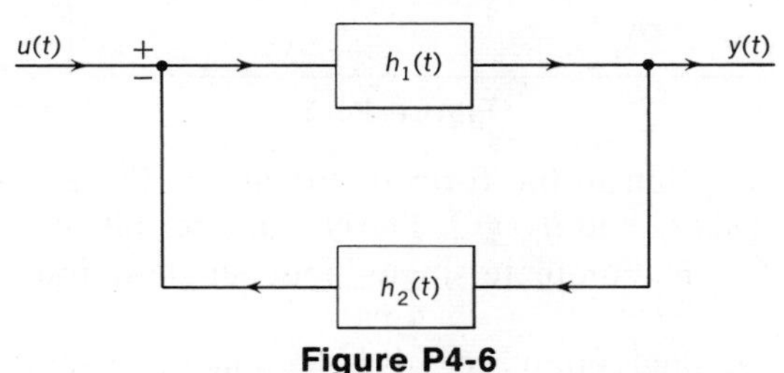

Figure P4-6

4-6 For the system represented by the block diagram of Figure P4-6

$$h_1(t) = \begin{cases} 0 & \text{for } t < 0 \\ e^{-(1+j20)t} + e^{-(1-j20)t} & \text{for } t \geq 0 \end{cases}$$

$$h_2(t) = \begin{cases} 0 & \text{for } t < 0 \\ Ke^{-\alpha t} & \text{for } t \geq 0 \end{cases}$$

Select values of K and α so that all natural frequencies are real.

4-7 Consider a single-input–two-output system characterized by equations

$$\dot{x} = -x + v$$

$$y_1 = x$$

$$y_2 = v - x$$

Compute the over-all $\underset{\sim}{A}$, $\underset{\sim}{B}$, $\underset{\sim}{C}$, and $\underset{\sim}{D}$ matrixes and the two transfer functions when: (a) the low-pass output y_1 is fed back through a gain K [Figure P4-7(a)]; (b) the high-pass output y_2 is fed back through a gain K [Figure P4-7(b)].

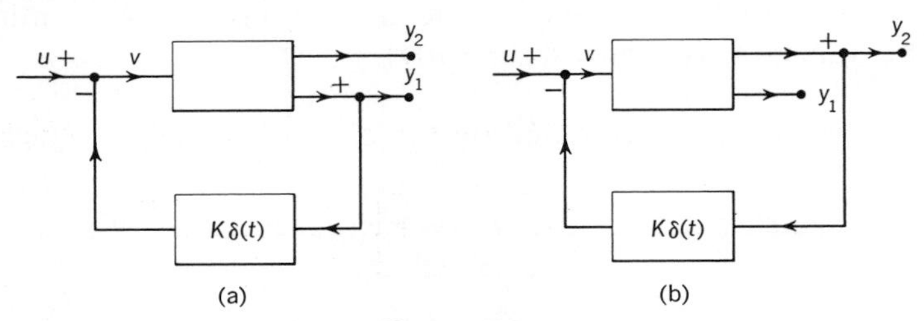

Figure P4-7

4-8 Consider the two-loop feedback system of Figure P4-8. Assume

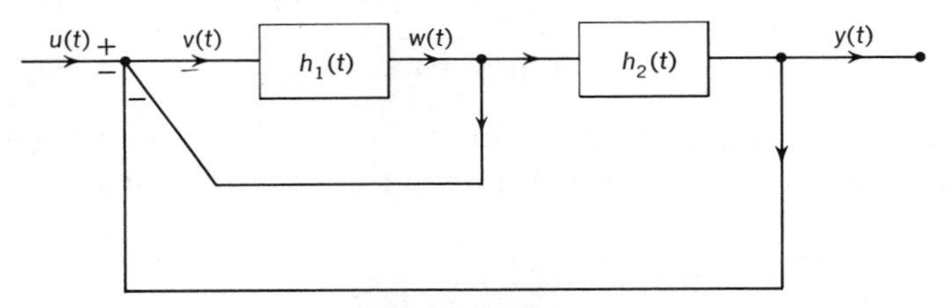

Figure P4-8

you have a description in the form of Equations (3-35) for each of the separate boxes [$h_1(t)$ and $h_2(t)$]. Derive an over-all set of equations in the same form for the complete single-input–single-output system.

4-9 Consider a single-loop negative feedback system with transfer function $H_1(j\omega)$ in the forward path and $H_2(j\omega)$ in the feedback path. The over-all transfer function is

$$\frac{H_1(j\omega)}{1 + H_1(j\omega)\, H_2(j\omega)}$$

Let $H_1(j\omega)$ have a second-degree denominator and $H_2(j\omega)$ a first-degree denominator. In most cases the over-all transfer function will have a third-degree denominator. Choose the constants so that there is cancellation and thus the over-all transfer function has only a second-degree denominator. For the values chosen, set up the separate $\underset{\sim}{A}$, $\underset{\sim}{B}$, $\underset{\sim}{C}$, and

$\underset{\sim}{D}$ matrixes and combine to give an over-all $\underset{\sim}{A}, \underset{\sim}{B}, \underset{\sim}{C}, \underset{\sim}{D}$ matrix description. Show that the cancelled natural frequency is still present in the $\underset{\sim}{A}$ matrix. State whether the system is degenerate because it is not controllable or not observable, or both.

4-10 Starting from Equations (3-35) show that the transfer function of single-input–single-output system is

$$H(j\omega) = \underset{\sim}{C}[j\omega \underset{\sim}{I} - \underset{\sim}{A}]^{-1} \underset{\sim}{B} + D + j\omega E$$

Repeat the derivation using Equations (4-78) and show that the matrix $\underset{\sim}{K}$ drops out and the same transfer function results provided $\underset{\sim}{K}$ and $\underset{\sim}{K}^{-1}$ both exist. You will need the fact that $\underset{\sim}{K}^{-1} \underset{\sim}{I} = \underset{\sim}{I} \underset{\sim}{K}^{-1}$; that is, the identity commutes with any square matrix. Thus, you have shown that the transfer function is independent of the choice of state variables provided the new state variables are related to the old by a nonsingular transformation.

4-11 Consider the system with two state variables and one input described by the equation

$$\dot{\mathbf{x}} = \begin{bmatrix} -1 & 0 \\ 0 & -2 \end{bmatrix} \mathbf{x} + \begin{bmatrix} 1 \\ 1 \end{bmatrix} v$$

Let $\mathbf{x}(0) = \begin{bmatrix} 2 \\ 1 \end{bmatrix}$. Find an input $v(t)$ on the interval $(0, 1)$ such that $\mathbf{x}(1) = \begin{bmatrix} 1 \\ 2 \end{bmatrix}$.

4-12 Consider the system with two state variables and one input described by the equation

$$\dot{\mathbf{x}} = \begin{bmatrix} -4 & 1 \\ -6 & 1 \end{bmatrix} \mathbf{x} + \begin{bmatrix} 3 \\ 7 \end{bmatrix} v$$

Let $\mathbf{x}(0) = \begin{bmatrix} 4 \\ 10 \end{bmatrix}$. Find an input $v(t)$ on the interval $(0, 1)$ such that $\mathbf{x}(1) = \begin{bmatrix} 5 \\ 11 \end{bmatrix}$.

4-13 Consider the system with two state variables and one output described by the equations

$$\dot{\mathbf{x}} = \begin{bmatrix} -4 & 1 \\ -6 & 1 \end{bmatrix} \mathbf{x}$$

$$y = [-\tfrac{1}{2} \;\; \tfrac{1}{2}]\, \mathbf{x}$$

Find $\mathbf{x}(0)$ if $y(0) = 0, y(1) = 1$.

4-14 Consider the single-input–single-output system described by

$$\dot{\mathbf{x}} = \begin{bmatrix} -4 & 1 \\ -6 & 1 \end{bmatrix} \mathbf{x} + \begin{bmatrix} 3 \\ 7 \end{bmatrix} v$$

$$y = \begin{bmatrix} -\frac{1}{2} & \frac{1}{2} \end{bmatrix} \mathbf{x} + v$$

Connect this system in the negative feedback configuration shown in Figure P4-14. Suppose $\mathbf{x}(0) = \begin{bmatrix} 1 \\ 0 \end{bmatrix}$. Draw a new block diagram good

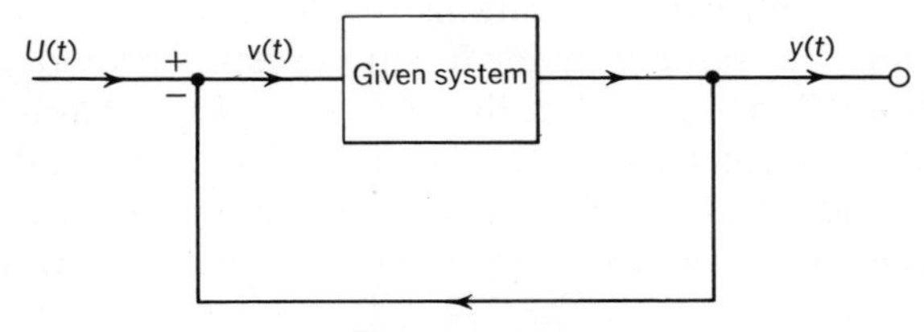

Figure P4-14

for $t > 0$ which accounts for these initial conditions as a second input. Write the over-all state variable equations for this new block diagram and verify that this two-input system with the state at $t = 0$ equal to zero has the same total response as the original system for $t > 0$.

CHAPTER

5

Discrete Time Signal Processing

In many engineering applications information is transmitted on a discrete time basis. That is, the important quantity is a number associated with an interval of time. The signals used to transmit discrete time information are called *discrete time signals*. In each time interval the signal is characterized by only one number. The most common discrete time signals are rectangular pulses. The characterizing number may be the pulse amplitude, the pulse duration or the pulse position within the time interval. For this important special class of signals there are special mathematical techniques for analyzing the signal processing by lumped systems. These techniques are the primary topics of discussion of this chapter.

5-1 Time-Domain Input-Output Relations for Lumped Systems

When a discrete time signal consisting of rectangular pulses is passed through a lumped, linear, time-invariant system, it is no longer a train of pulses when it emerges. When a large system is operating on discrete time signals, each lumped system section must be followed by a modulator to convert its output back to discrete time form. To be specific, consider the system shown in Figure 5-1. Let us now see how this system processes the various types of discrete time signals mentioned in the previous paragraph.

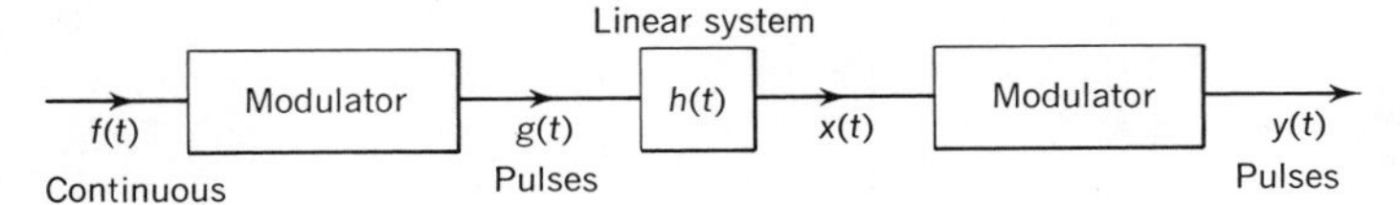

Figure 5-1 Basic system for discrete time signals.

Sample and hold modulators. For the first case let the modulators in Figure 5-1 be a synchronized sample and hold devices with sampling period T $(S + H - T)$. Then

$$g(t) = f(kT),\ kT \leqslant t < (k+1)\,T,\ k \text{ an integer} \tag{5-1}$$

$$x(t) = g * h = \dot{g} * \underset{\cdot}{h} \tag{5-2}$$

provided the convolution, $\dot{g}$, and $\underset{\cdot}{h}$ exist at least in the sense of the previous chapter. But $\dot{g}(t)$ is a string of delta functions

$$\dot{g}(t) = \sum_{k=-\infty}^{\infty} [f(kT) - f((k-1)T)]\ \delta(t - kT) \tag{5-3}$$

Then

$$x(t) = \sum_{k=-\infty}^{\infty} [f(kT) - f((k-1)T)]\ \underset{\cdot}{h}(t - kT) \tag{5-4}$$

The expression (5-4) is somewhat easier to work with if the summation index is modified as follows: We can rewrite

$$\sum_{k=-\infty}^{\infty} f((k-1)T)\ \underset{\cdot}{h}(t - kT)$$

as

$$\sum_{k=-\infty}^{\infty} f(kT)\ \underset{\cdot}{h}(t - (k+1)T)$$

All terms are still included. With this change (5-4) becomes

$$x(t) = \sum_{k=-\infty}^{\infty} f(kT)\ [\underset{\cdot}{h}(t - kT) - \underset{\cdot}{h}(t - (k+1)T)] \tag{5-5}$$

Multiplying and dividing by T gives

$$x(t) = \sum_{k=-\infty}^{\infty} Tf(kT) \left[\frac{\underset{\cdot}{h}(t - kT) - \underset{\cdot}{h}(t - (k+1)T)}{T}\right]$$

If h is sufficiently smooth and T sufficiently small, the bracket is a good approximation to the derivative of $\underset{\cdot}{h}$. Then

$$x(t) \approx x_a(t) = \sum_{k=-\infty}^{\infty} Tf(kT)\, h(t-kT) \tag{5-6}$$

where the subscript a is inserted to indicate approximate. Since the evaluation of (5-6) requires only half the computation of (5-5), this last expression is frequently used. It is valid when the sampling period is short compared to the time constants of the linear, time-invariant system.

The expressions (5-5) and (5-6) are infinite sums whose convergence must be assured before they can be used as the values for $x(t)$. Since (5-5) is exactly (5-2), the latter is well-defined if the former is. The sum (5-6) is a discrete approximation to the convolution of $f(t)$, the input to the modulator in Figure 5-1, with the response function $h(t)$. Thus if $h(t)$ is stable in the sense that bounded inputs give bounded outputs and if $f(t)$ is bounded, then the sum is well-defined when the sampling period T is sufficiently small. For certain special but important cases, these nebulous convergence conditions will be made more precise.

When the system characterized by $h(t)$ in Figure 5-1 is causal, $h(t)$ is zero for $t < 0$. In this case the terms in the infinite sums (5-5) and (5-6) are zero for k greater than the largest integer in t/T. Using the symbol $[t/T]$ for this largest integer, (5-5) and (5-6) become

$$x(t) = \sum_{k=-\infty}^{[t/T]} f(kT)\, [\underset{\cdot}{h}(t-kT) - \underset{\cdot}{h}(t-(k+1)T)] \tag{5-7}$$

$$x_a(t) = \sum_{k=-\infty}^{[t/T]} Tf(kT)\, h(t-kT) \tag{5-8}$$

When $h(t)$ represents a lumped, linear, time-invariant system, which is also described by state variable equations of the form (3-35), then some simply stated conditions guarantee the convergence of the sums in (5-7) and (5-8). If the response function $h(t)$ contains only exponentials and polynomials times exponentials,[1] if the natural frequencies all have negative real parts, and if the input $f(t)$ is bounded, then the sums converge.

In the next section a technique for summing the series is discussed. Before taking up this technique let us set up the response of systems shown in Figure 5-1 with other types of modulators.

Pulse amplitude modulators. If the modulator in Figure 5-1 is a pulse amplitude modulator that generates pulses of length Δ and height $f(kT)$, then

$$g(t) = \begin{cases} f(kT) & \text{for } kT \leqslant t \leqslant kT + \Delta \\ 0 & \text{for } (kT+\Delta) < t < (k+1)T \end{cases} \tag{5-9}$$

[1] In terms of Equation (3-35) $\underset{\sim}{D}$ and $\underset{\sim}{E}$ must be zero.

where k takes on all integer values, and $\Delta < T$.

Now

$$\dot{g}(t) = \sum_{k=-\infty}^{\infty} f(kT)\,[\delta(t-kT) - \delta(t-kT-\Delta)] \tag{5-10}$$

Then

$$x(t) = \sum_{k=-\infty}^{\infty} f(kT)\,[\underset{\cdot}{h}(t-kT) - \underset{\cdot}{h}(t-kT-\Delta)] \tag{5-11}$$

As with the sample and hold modulator a simpler approximate formula for $x(t)$ can be used if Δ is small compared to variations of $\underset{\cdot}{h}(t)$. That is, if

$$h(t-kT) \approx \frac{\underset{\cdot}{h}(t-kT) - \underset{\cdot}{h}(t-kT-\Delta)}{\Delta} \tag{5-12}$$

then it is often reasonable to approximate (5-11) by

$$x(t) \approx x_a(t) = \sum_{k=-\infty}^{\infty} \Delta\, f(kT)\, h(t-kT) \tag{5-13}$$

We cannot always be sure that if Equation (5-12) is a good approximation for all k, then the sum (5-13) is also a good approximation. The small errors in each term may add to a big error in the sum. Each case must be examined separately. The techniques of the next section can often be used to justify the approximation of (5-13).

Pulse width modulators. For a pulse width modulator

$$g(t) = \sum_{k=-\infty}^{\infty} [U(t-kT) - U(t-(k+af(kT))T)] \tag{5-14}$$

where the constant a must be selected so that $[af(kT)] < 1$ and $f(t)$ must be biased to be positive. Since

$$\dot{g}(t) = \sum_{k=-\infty}^{\infty} [\delta(t-kT) - \delta(t-(k+af(kT))T)] \tag{5-15}$$

$$x(t) = \sum_{k=-\infty}^{\infty} [\underset{\cdot}{h}(t-kT) - \underset{\cdot}{h}(t-(k+af(kT))T)] \tag{5-16}$$

As with the sample and hold modulator and the pulse amplitude modulator, a simpler approximate formula for $x(t)$ can be obtained from (5-16) by approximating the difference by a derivative. To make the bracket in (5-16) into a difference quotient requires a $Tf(kT)$ in the denominator. To this end (5-16) can be rewritten as

$$x(t) = \sum_{k=-\infty}^{\infty} aTf(kT)\, \frac{\underset{\cdot}{h}(t-kT) - \underset{\cdot}{h}(t-kT-aTf(kT))}{aTf(kT)} \tag{5-17}$$

If $[aT\, f(kT)]$ is always small compared to the variation of $h(t)$, then the difference quotient is a good approximation to the derivative. Thus

$$x(t) \approx x_a(t) = \sum_{k=-\infty}^{\infty} aTf(kT)\, h(t-kT) \tag{5-18}$$

Note that $aTf(kT)$ is the area of the kth pulse. In fact, all the approximate formulas (5-6), (5-13), and (5-18) have the form $\Sigma_{k=-\infty}^{\infty} A_k h(t-kT)$ where A_k is the area of the kth pulse. This approximate formula can be summed by the methods of the next section whereas (5-16) cannot. An expression for an upper bound on the error can be derived in many cases as shown in Section 5-3.

5-2 Summation of Discrete Convolutions—The Z-Transform

In the previous section the analysis problem for the system of Figure 5-1 was carried to the point where a particular summation had to be computed. All the sums had the form

$$w(t) = \sum_{k=-\infty}^{\infty} p(t-kT)\, q(kT) \tag{5-19}$$

We shall refer to (5-19) as a discrete convolution.

With many well-behaved functions, discrete convolutions are as easy to perform as convolution integrals. For example, suppose

$$p(t) = \begin{cases} 0 & \text{for } t < 0 \\ e^{-at} & \text{for } t \geqslant 0;\ a \text{ real and positive} \end{cases}$$

$$q(t) = U(t), \text{ the unit step}$$

Then at the sampling points nT

$$\begin{aligned} w(nT) &= \sum_{k=0}^{n} e^{-a(n-k)T} \\ &= \sum_{i=0}^{n} (e^{-aT})^i \end{aligned} \tag{5-20}$$

Since a and T are both positive, e^{-aT} is less than one. Thus

$$\sum_{i=0}^{\infty} (e^{-aT})^i = \frac{1}{1-e^{-aT}} \tag{5-21}$$

Now Equation (5-20) can be written as

$$w(nT) = \sum_{i=0}^{\infty} (e^{-aT})^i - \sum_{i=n+1}^{\infty} (e^{-aT})^i$$

$$= \sum_{i=0}^{\infty} (e^{-aT})^i \, [1 - e^{-(n+1)aT}] \tag{5-22}$$

$$= \frac{1 - e^{-(n+1)aT}}{1 - e^{-aT}}$$

This last expression is the desired closed form for $w(nT)$ for $n \geqslant 0$.

The Z-transform defined. The idea of using the geometric series (5-21) can be extended to aid in closed form summation of many discrete convolutions through a transform—the Z-transform. Given a function f such that $f(t)$ is defined for all real t, than the Z-transform is defined by

$$F_T(z) = \sum_{k=-\infty}^{\infty} f(kT) z^{-k} \tag{5-23}$$

Clearly the Z-transform of any function is well-defined for only certain ranges of values of z. For some functions there are no values of z such that the series (5-23) converges absolutely. We say then the Z-transform of such a function does not exist.

To show the application of Z-transforms to discrete convolutions we consider the product of two Z-transforms. That is, given functions f and g such that their Z-transforms both exist for some z, we consider

$$\begin{aligned} F_T(z)\, G_T(z) &= \sum_{k=-\infty}^{\infty} f(kT)\, z^{-k} \sum_{k=-\infty}^{\infty} g(mT)\, z^{-m} \\ &= \sum_{k=-\infty}^{\infty} \sum_{m= \infty}^{\infty} f(kT)\, g(mT)\, z^{-(k+m)} \\ &= \sum_{k=-\infty}^{\infty} \sum_{p=-\infty}^{\infty} f(kT)\, g((p-k)T)\, z^{-p} \\ &= \sum_{p=-\infty}^{\infty} \left[\sum_{k=-\infty}^{\infty} f(kT)\, g((p-k)T) \right] z^{-p} \end{aligned} \tag{5-24}$$

Thus the Z-transform of a discrete convolution is the product of the Z-transforms of the two functions being convolved, provided that the transforms of both exist for a particular value of z. The requirement for absolute convergence for the existence of the Z-transform allows the interchange of order of summation in (5-24).

With a table of Z-transforms we can work with discrete time signal processing just as we use Laplace transforms in lumped systems with continuous input signals. Two of the more important functions in discrete time systems are the unit step $U(t)$, and the exponential that starts at $t=0$. If

$$f(t) = U(t) \qquad F_T(z) = \sum_{k=0}^{\infty} z^{-k} = \frac{z}{z-1} \qquad \text{for } |z| > 1 \tag{5-25}$$

If for $t < 0, f(t) = 0$ and for $t \geqslant 0$

$$f(t) = e^{\alpha t} \qquad F_T(z) = \sum_{k=0}^{\infty} e^{k\alpha T} z^{-k} = \sum_{k=0}^{\infty} (e^{\ln z - \alpha T})^{-k}$$
$$= \frac{ze^{-\alpha T}}{ze^{-\alpha T} - 1} \qquad \text{for } |z| > e^{\alpha T} \tag{5-26}$$

Inverse Z-transforms. With (5-25) and (5-26) as a table of Z-transforms we can compute $w(nT)$ of (5-20) as follows:

$$W(z) = \left[\frac{z}{z-1}\right]\left[\frac{z}{z-e^{-aT}}\right] \tag{5-27}$$

Formula (5-26) is valid for $|z| > e^{-aT}$ and (5-25) for $|z| > 1$ in this case. Thus (5-27) is valid for $|z| > 1$. The Z-transform, like the Laplace transform, is a linear operator. Thus the Z-transform of the sum is the sum of the Z-transforms. This fact allows us to use partial fractions to find the inverse Z-transform.

Since both transforms in our table have z as a numerator factor, we can most easily find a transform in the table if this factor is present. To this end we expand (5-27) as follows:

$$W(z) = z\left[\frac{z}{(z-1)(z-e^{-aT})}\right]$$
$$= z\left[\frac{\dfrac{1}{1-e^{-aT}}}{z-1} + \frac{\dfrac{e^{-aT}}{e^{-aT}-1}}{z-e^{-aT}}\right] \tag{5-28}$$

Some manipulation makes the two terms readily recognizable from the table as follows:

$$W(z) = \frac{1}{1-e^{-aT}}\left[\frac{z}{z-1}\right] + \frac{e^{-aT}}{e^{-aT}-1}\left[\frac{z}{z-e^{-aT}}\right] \tag{5-29}$$

Thus, for $t \geqslant 0$,

$$w(t) = \frac{1}{1-e^{-aT}}\,[1 - e^{-aT}\,e^{-at}] \tag{5-30}$$

At the sampling points for $n \geqslant 0$

$$w(nt) = \frac{1 - e^{-a(n+1)T}}{1 - e^{-aT}} \tag{5-31}$$

We note that Equation (5-31) is identical to (5-22).

5-3 Z-Transform Analysis of Lumped Systems

In order to apply Z-transform analysis to expressions such as (5-5) we need a shifting theorem. To derive the necessary formula we examine the transform of $f(t-nT)$. Using the notation $Z_T(f(t))$ for the Z-transform we have

$$\begin{aligned} Z_T(f(t-nT)) &= \sum_{k=-\infty}^{\infty} f(kT-nT)\ z^{-k} \\ &= \sum_{p=-\infty}^{\infty} f(pT)\ z^{-(p+n)} \\ &= z^{-n}\ Z_T(f(t)) \end{aligned} \tag{5-32}$$

As an example of the application of Z-transforms to a system problem let us consider the sample and hold modulator problem whose response is characterized by (5-5). Let

$$h(t) = \begin{cases} e^{pt} + e^{p^*t} & \text{for } t \geqslant 0 \\ \quad 0 & \text{otherwise} \end{cases} \tag{5-33}$$

$$f(t) = \begin{cases} 1 & \text{for } 0 \leqslant t \leqslant 5 \\ 0 & \text{otherwise} \end{cases} \tag{5-34}$$

$$T = 1$$

Then for $t \geqslant 0$

$$\underset{\cdot}{h}(t) = \frac{1}{p}\ e^{pt} + \frac{1}{p^*}\ e^{p^*t} - 2\ \text{Re}\left[\frac{1}{p}\right] \tag{5-35}$$

The appropriate Z-transforms are

$$Z_1(h(t)) = \frac{z}{z-e^p} + \frac{z}{z-e^{p^*}} \tag{5-36}$$

$$Z_1(\underset{\cdot}{h}(t)) = \frac{(1/p)\ z}{z-e^p} + \frac{(1/p^*)\ z}{z-e^{p^*}} - \frac{2\ \text{Re}\ [1/p]\ z}{z-1} \tag{5-37}$$

$$\begin{aligned} Z_1(f(t)) &= 1 + z^{-1} + z^{-2} + z^{-3} + z^{-4} + z^{-5} \\ &= \frac{z(1-z^{-6})}{z-1} \end{aligned} \tag{5-38}$$

Now the Z-transform of the output is easily found from the above formulas and the shifting rule (5-32). It is

$$\begin{aligned} Z_1(x(t)) &= \left(\frac{z(1-z^{-6})}{z-1}\right)\ (1-z^{-1})\ Z_1\ (\underset{\cdot}{h}(t)) \\ &= (1-z^{-6})\ [Z_1(\underset{\cdot}{h}(t))] \end{aligned} \tag{5-39}$$

With the aid of the shifting rule, the inverse transform of Equation (5-39) is immediately recognized as

$$x(t) = \begin{cases} 0 & \text{for } t < 0 \\ \dfrac{1}{p} e^{pt} + \dfrac{1}{p^*} e^{p^*t} - 2 \operatorname{Re}\left[\dfrac{1}{p}\right] & \text{for } 0 \leqslant t \leqslant 6 \\ \dfrac{1}{p}(1 - e^{-6p})\, e^{pt} + \dfrac{1}{p^*}(1 - e^{-6p^*})\, e^{p^*t} & \text{for } t > 6 \end{cases} \tag{5-40}$$

This final result should have been obvious since the sample and hold device would present a step of duration 6 s to the lumped linear system in this case.

Z-transforms of the approximate response Let us now consider the same example and try the approximate formula (5-6). From Equations (5-36) and (5-38), the Z-transform of $x_a(t)$ is

$$\begin{aligned} Z_1(x_a(t)) &= \frac{z(1 - z^{-6})}{z - 1}\left[\frac{z}{z - e^p} + \frac{z}{z - e^{p^*}}\right] \\ &= z(1 - z^{-6})\left[\frac{\dfrac{e^p}{e^{-p} - 1} + \dfrac{e^{-p^*}}{e^{-p^*} - 1}}{z - 1} \right. \\ &\qquad \left. + \frac{\dfrac{e^p}{e^p - 1}}{z - e^p} + \frac{\dfrac{e^{p^*}}{e^{p^*} - 1}}{z - e^{p^*}}\right] \end{aligned} \tag{5-41}$$

From the last line partial fraction expansion in (5-41) we immediately recognize the inverse transform:

$$x_a(t) = \begin{cases} 0 & \text{for } t < 0 \\ \dfrac{1}{1 - e^p} + \dfrac{1}{1 - e^{p^*}} + \dfrac{e^{pt}}{1 - e^{-p}} + \dfrac{e^{p^*t}}{1 - e^{-p}} & \text{for } 0 \leqslant t \leqslant 6 \\ \dfrac{(1 - e^{-6p})\, e^{pt}}{1 - e^{-p}} + \dfrac{(1 - e^{-6p^*})\, e^{p^*t}}{1 - e^{-p^*}} & \text{for } 6 \leqslant t \end{cases} \tag{5-42}$$

A comparison of $x_a(t)$ in (5-42) with the exact expression, $x(t)$ in (5-40) shows that the approximation is good if $|p| \ll 1$. Specifically the only difference is that p in (5-40) is replaced by $1 - e^{-p}$ in (5-42). The power series for the exponential shows the approximation.

Errors in the approximate formulas For all the modulators discussed above the approximate formulas are good for a lumped system when the natural frequencies are all small in magnitude compared to the sampling frequency. To prove this statement let us begin with one term of the form

e^{pt} in $h(t)$ and examine each of the approximate formulas given above. For each case

$$h(t) = e^{pt} \qquad \text{for } t \geqslant 0$$

$$\underset{\cdot}{h}(t) = \frac{1}{p}\,(e^{pt} - 1) \quad \text{for } t \geqslant 0$$

For the sample and hold case the difference between (5-5) and (5-6) gives the error at nT as

$$\epsilon_{S+H}(nT) = \sum_{k=-\infty}^{n} f(kT)\left[\frac{1-e^{-pT}}{p} - T\right] e^{p(n-k)T} + f(nT)\left[\frac{e^{-pT}-1}{p}\right] \tag{5-43}$$

The last term in (5-43) must be added since $h(t)$ is zero for $t < 0$ and formula (5-5) picks up $\underset{\cdot}{h}(-T)$ when the summation is taken from $-\infty$ to $k = n$. We expect such a term to appear in the error since the approximate formula (5-6) would show an output at nT due to the sample taken at nT if $h(0) \neq 0$. Actually the sample taken at nT does not effect the response until after the sampling instant.

A series expansion of the exponentials in (5-43) gives

$$\epsilon_{S+H}(nT) = \left[-\frac{(pT)^2}{2p} + \frac{(pT)^3}{3!\,p} - \cdots\right] \sum_{k=-\infty}^{n} f(kT)\, e^{p(n-k)T} + f(nT)\left[-T + \frac{(pT)^2}{2p} + \frac{(pT)^3}{3!\,p} + \cdots\right] \tag{5-44}$$

If $|pT| \ll 1$, the term in the bracket in front of the summation is small. Furthermore if $f(t)$ is bounded and Re $[p] < 0$, then the sum converges absolutely. Thus except for the term $-Tf(nT)$, the error goes to zero as pT goes to zero. It is easy to show that if $h(0) = 0$, the contribution to the error from other exponentials will cancel the $-Tf(nT)$ term (see Problem 5-1).

For the pulse amplitude modulator, the error in the approximate analysis is obtained by subtracting (5-13) from (5-11). The result is

$$\begin{aligned} \epsilon_{\text{PAM}} &= \frac{1 - p\,\Delta - e^{-p\Delta}}{p} \sum_{k=-\infty}^{n} f(kT)\, e^{p(n-k)T} + f(nT)\left[\frac{e^{-p\Delta}-1}{p}\right] \\ &= \left[-\frac{(p\,\Delta)^2}{2p} + \frac{(p\,\Delta)^3}{3!\,p} - \cdots\right] \sum_{k=-\infty}^{n} f(kT)\, e^{p(n-k)T} + f(nT)\left[-\Delta + \frac{(p\Delta)^2}{2p} - \cdots\right] \end{aligned} \tag{5-45}$$

Comparison of (5-43) and (5-44) shows that so far as the error is concerned, the $S + H$ modulator is a special case of the PAM modulator with pulse width equal to the sampling period.

For the pulse width modulator the error is obtained by subtracting (5-18) from (5-16). For a single exponential in $h(t)$ the contribution to the error is

$$\begin{aligned}\epsilon_{\text{PWM}} &= \sum_{k=-\infty}^{n} e^{p(n-k)T}\left[\frac{1-e^{-paTf(kT)}-paTf(kT)}{p}\right] \\ &\quad + \frac{e^{-patf(nT)}-1}{p} \\ &= \sum_{k=-\infty}^{n}\left[-\frac{[paTf(kT)]^2}{2p}+\frac{[paTf(kT)]^3}{3!\,p}+\cdots\right]e^{p(n-k)T} \\ &\quad + \left[aTf(nT)-\frac{(paTf(nT))^2}{2p}+\cdots\right]\end{aligned} \tag{5-46}$$

In the definition (5-14) of a pulse width modulator we had to restrict $0 \leqslant af(kT) < 1$ to prevent saturation. With this condition, (5-46) goes to zero as pT goes to zero except for the term $aTf(nT)$. As with the previous modulators considered, this term is cancelled when $h(0) = 0$. The above error formulas are for a single exponential term in the response function $h(t)$. Similar formulas can be derived for a term of the form $t^m e^{pt}$ (see Problem 5-2). From these error formulas we can make the following statement:

> For the system of Figure 5-1, with the three types of modulators discussed in Section 5-1, the approximate analysis is valid when the linear system is lumped, the product of the highest natural frequency of the system with the sampling period is much less than one, and the response function is zero for $t = 0$. Now that we know when the approximate formulas are valid, let us tabulate all the formulas for the output in both the time and Z-transform domains. These are given in Table 5-1.

Z-transforms of interconnected systems. The real advantage of the Z-transform is in the analysis of system interconnections. To show this let us consider the simple feedback circuit of Figure 5-2. To be specific let

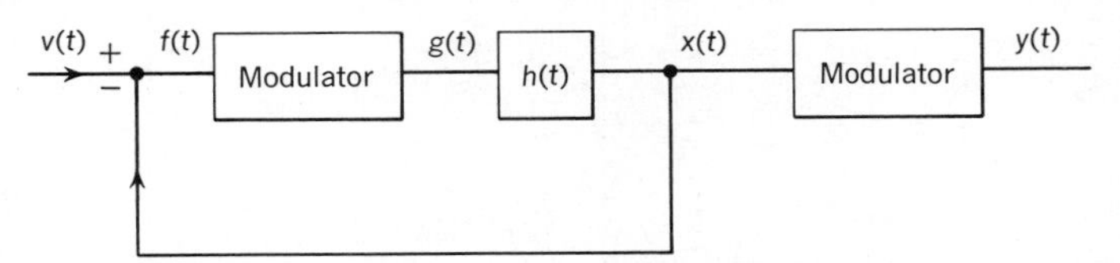

Figure 5-2 Discrete time signal processor with feedback.

$$h(t) = \begin{cases} 0 & \text{for } t < 0 \\ e^{pt} & \text{for } t \geqslant 0 \end{cases} \tag{5-47}$$

Table 5-1

MOD	$g(t)$	$x(t)$	$x_a(t)$	$X(z)$	$x_a(z)$	ϵ
$S+H$	$f(kT)$ for $kT \leq t < (k+1)T$	$\sum_{k=-\infty}^{\infty} f(kT)[\underset{\cdot}{h}(t-kT) - \underset{\cdot}{h}(t-k+1)T)$	$\sum_{k=-\infty}^{\infty} Tf(kT)h(t-kT)$	$F(z)\left(\frac{z-1}{z}\right) Z_T(\underset{\cdot}{h}(t))$	$TF(z)H(z)$	(5-44)
PAM	$f(kT)$ for $kT \leq t < kT+\Delta$ 0 for $(kT+\Delta) < t < (k+1)T$ $\Delta < T$	$\sum_{k=-\infty}^{\infty} f(kT)[\underset{\cdot}{h}(t-kT) - \underset{\cdot}{h}(t-kT-\Delta)]$	$\sum_{k=-\infty}^{\infty} \Delta f(kT)h(t-kT)$	No simple formula	$\Delta F(z)H(z)$	(5-45)
PWM	$\sum_{k=-\infty}^{\infty} [U(t-kT) - U(t(k+af(kT))T)]$ $0 \leq af(kT) < 1$	$\sum_{k=-\infty}^{\infty} [\underset{\cdot}{h}(t-kT) - \underset{\cdot}{h}(t-(k+af(kT))T)]$	$\sum_{k=-\infty}^{\infty} aTf(kT)h(t-kT)$	No simple formula	$aTF(z)H(z)$	(5-46)
PPM	$\sum_{k=-\infty}^{\infty} [U(t-kT-af(kT)) - U(t-kT-af(kT)-\Delta)]$ $0 \leq af(kT) < T-\Delta$	See Problem 5-3	See Problem 5-3	No simple formula	See Problem 5-3	See Problem 5-3

$$v(t) = \begin{cases} \cos \omega t & \text{for } 0 \leqslant t \leqslant 5 \\ 0 & \text{otherwise} \end{cases} \tag{5-48}$$

$$T = 1$$

Then

$$\underset{\cdot}{h}(t) = \begin{cases} 0 & \text{for } t < 0 \\ \dfrac{1}{p} e^{pt} - \dfrac{1}{p} & \text{for } t \geqslant 0 \end{cases} \tag{5-49}$$

The time-domain equations for this system are

$$\begin{aligned} x(t) &= \sum_{k=-\infty}^{\infty} f(kT)\,[\underset{\cdot}{h}(t - kT) - \underset{\cdot}{h}(t - (k+1)T] \\ f(t) &= v(t) - x(t) \end{aligned} \tag{5-50}$$

When these equations are Z-transformed they become linear algebraic equations with $Z_1(x(t))$ as the unknown. They can be solved easily. The Z_1-transform of $\underset{\cdot}{h}(t)$ is

$$Z_1(\underset{\cdot}{h}(t)) = \frac{1}{p}\left[\frac{z}{z - e^p} - \frac{z}{z-1}\right] \tag{5-51}$$

The Z_1-transform of $v(t)$ is

$$V(z) = \frac{1 - z^{-5}}{2}\left[\frac{z}{z - e^{j\omega}} + \frac{z}{z - e^{-j\omega}}\right] \tag{5-52}$$

Thus the Z-transform of Equation (5-50) is

$$X(z) = V(z)\left(\frac{z-1}{z}\right) Z_1(\underset{\cdot}{h}(t)) - X(z)\left(\frac{z-1}{z}\right) Z_1(\underset{\cdot}{h}(t)) \tag{5-53}$$

Then

$$\begin{aligned} X(z) &= \frac{\left(\dfrac{z-1}{z}\right) V(z)\, Z_1(\underset{\cdot}{h}(t))}{1 + \left(\dfrac{z-1}{z}\right) Z_1(\underset{\cdot}{h}(t))} \\ &= (1 - z^{-5}) \frac{\dfrac{(z-1)[2z^2 - z(e^{j\omega} + e^{-j\omega})][ze^p - z]}{zp(z - e^{j\omega})(z - e^{-j\omega})(z - e^p)(z-1)}}{\dfrac{pz(z - e^p)(z-1) + (z-1)(ze^p - z)}{pz(z - e^p)(z-1)}} \\ &= \frac{(1 - z^{-5})[2z - (e^{j\omega} + e^{-j\omega})]\, z(e^p - 1)}{(z - e^{j\omega})(z - e^{-j\omega})(pz + e^p(1-p) - 1)} \end{aligned} \tag{5-54}$$

The last line of (5-54) is readily expandable in partial fractions in such a way that the corresponding time function is easily recognized.[2]

[2] See Equation (5-29) above for the method and Problem 5-4 for details.

For the other types of modulator the Z-transform does not apply so easily to the exact analysis. When the approximate formulas are valid, then the Z-transform is applicable in exactly the way it is used above.[3] Analysis of other systems and estimate of the errors involved are left to the problems.

5-4 Linear Discrete Time Systems

In the previous sections of this chapter we discussed processing of discrete time signals by lumped systems. Often discrete time signals are processed by systems that are themselves discrete. That is, the values of the system variables are all discrete time signals. The values of these variables in one interval depend on their values in the previous interval and also on the values of the discrete time signal inputs to the system in the previous interval. Although such systems are not lumped by the definition in Chapter 1, the method of analysis is sufficiently similar to the analysis methods for lumped systems to make their inclusion justified.

Solution of a first-order difference equation. Specifically a one-variable linear, time-invariant discrete time system is described by the difference equation

$$x((k+1)T) = ax(kT) + v(kT) \tag{5-55}$$

where x is the system variable; T is the interval length; k is an integer; a is the constant of the system; $v(kT)$ is the input in the kth interval.

A linear difference equation can be solved by variation of parameters just as the differential equations were above. Thus for (5-55) the first step is to solve the homogeneous equation

$$y((k+1)T) = ay(kT) \tag{5-56}$$

The functional form that satisfies the general homogeneous linear difference equation with constant coefficients is

$$y(kT) = Aq^{kT} \tag{5-57}$$

where A and q are constants.

Substituting into (5-56) gives

$$Aq^{(k+1)T} = aAq^{kT}$$

[3] When the final Z-transform cannot be put in a form that appears in available tables, there is an integral expression for the inverse transform (see Reference 28). This technique is not necessary for the analysis of the systems discussed in this text.

Dividing both sides by Aq^{kT} leaves

$$q^T = a$$

Thus

$$y(kT) = Aa^k \tag{5-58}$$

where A is constant determined by the initial conditions.

The variation of parameters step starts with the functional form (5-58) with A a function of kT and assumes a solution to Equation (5-55) as

$$x(kT) = A(kT)a^k \tag{5-59}$$

Substituting into Equation (5-55) gives

$$A((k+1)T)a^{(k+1)} = a\,A(kT)a^k + v(kT)$$

Regrouping gives

$$A((k+1)T) - A(kT) = a^{-(k+1)}\,v(kT) \tag{5-60}$$

A first-order difference equation such as (5-60) can be summed just as the first-order differential equation $\dot{x} = f(t)$ is integrated. For the equation

$$f((k+1)T - f(kT) = g(kT) \tag{5-61}$$

the solution is

$$f(kT) = \sum_{n=1}^{\infty} g((k-n)T) + K \tag{5-62a}$$

or

$$f(kT) = -\sum_{n=0}^{\infty} g((k+n)T) + K \tag{5-62b}$$

The choice between the two forms (5-62a) and (5-62b) depends on the convergence of the respective infinite series. The fact that both are solutions if the series converge is readily verified by substituting into (5-61). The constant K must be determined by an initial condition.

For a physical system the form (5-62a) is usually appropriate since it requires knowledge of past inputs [(5-62b) requires knowledge of future inputs], and the series converges if the past inputs are bounded and the system is stable.

Using Equation (5-62a) in (5-60) gives

$$A(kT) = \sum_{n=1}^{\infty} a^{-(k+1-n)}\,v((k-n)T) + K \tag{5-63}$$

Substituting into (5-59) gives the system response as

$$x(kT) = \sum_{n=1}^{\infty} \frac{1}{a} a^n v((k-n)T) + K\, a^k \tag{5-64}$$

If $x(kT)$ is known for one particular value of k (usually taken at $k = 0$ to simplify the notation), then the constant K is determined. This value of $x(0)$ is the state of the system at $k = 0$. The summation in (5-64) is the zero-state response, and the term $K\, a^k$ is the zero-input response. If $|a|$ is less than one, $K\, a^k \to 0$ as $k \to \infty$ so the system is stable. Furthermore, the sum in the zero-state response converges so long as the input $v(kT)$ is bounded as $k \to -\infty$.

The zero-state response term in (5-64) can be put in a form for computation via z-transforms. In this case the zero-state term is

$$x(kT) = \sum_{n=1}^{\infty} a^{n-1}\, v((k-n)T) \tag{5-65}$$

By a change of variables in the sum, (5-65) becomes

$$x(kT) = \sum_{m=0}^{\infty} a^m v((k-1-m)T) \tag{5-66}$$

The right side of (5-66) can now be summed using z-transforms.

Z-transforms for discrete systems. If we define a discrete system response function by

$$h(kT) = \begin{cases} a^k & \text{for } 0 \leqslant k \\ 0 & \text{for } k < 0 \end{cases} \tag{5-67}$$

Then the z-transform of the right side of (5-66) is the product of $H(z)$ with the z-transform of $v((k-1)T)$. By the shifting theorem (5-32) this latter transform is $(1/z)\ V(z)$. Now the z-transform of (5-66) is

$$X(z) = H(z)\frac{V(z)}{z} \tag{5-68}$$

From (5-67) and the definition of the z-transform

$$H(z) = \frac{z}{z-a}$$

Thus

$$X(z) = \frac{V(z)}{z-a}$$

If the input is a unit step, then (5-25) shows that

$$X(z) = \frac{z}{(z-a)(z-1)} = z\left[\frac{\frac{1}{a-1}}{z-a} + \frac{\frac{1}{1-a}}{z-1}\right]$$

Thus

$$x(kT) = \frac{1}{a-1} a^k + \frac{1}{1-a} U(t)$$

Similar results for sine wave inputs or other inputs with easily computed z-transforms are left to the problems.

Higher-order systems. The above results for first-order discrete systems generalize to high-order systems just as first-order lumped-system results generalized readily through normal form equations. In vector notation the normal form difference equations for a discrete system are

$$\mathbf{x}((k+1)T) = \underset{\sim}{A}\,\mathbf{x}(kT) + \underset{\sim}{B}\,\mathbf{y}(kT)$$
$$\mathbf{y}(kT) = \underset{\sim}{C}\,\mathbf{x}(kT) + \underset{\sim}{D}\,\mathbf{y}(kT) \tag{5-69}$$

Here the output $\mathbf{y}$ is taken as a linear combination of the state variables and the inputs all at the same time interval.

For the homogeneous part of Equation (5-69) we have

$$\mathbf{w}((k+1)T) = \underset{\sim}{A}\,\mathbf{w}(kT) \tag{5-70}$$

For a solution we assume

$$\mathbf{w} = q^{kT}\,\mathbf{W}$$

where q is a constant and $\mathbf{W}$ is a vector of constants, both to be determined. Now Equation (5-70) becomes

$$q^{(k+1)T}\,\mathbf{W} = q^{kT}\,\underset{\sim}{A}\,\mathbf{W}$$

Or, multiplying by q^{-kT} and regrouping,

$$[q^T \underset{\sim}{I} - \underset{\sim}{A}]\,\mathbf{W} = 0 \tag{5-71}$$

From (5-71) we see that the (q^T) are the eigenvalues of the $\underset{\sim}{A}$ matrix.

From the n eigenvalues of $\underset{\sim}{A}$ and the associated $\mathbf{W}$ vectors from (5-71), a fundamental matrix $\phi(kT)$ can be constructed. In the most general case $\underset{\sim}{\phi}(kT)$ has n^2 free constants just as in the differential equation case. That $\underset{\sim}{\phi}$ matrix that is the identity for $k = 0$ is given by the matrix expansion of $\underset{\sim}{A}^k$. The remainder of the derivation of the total response of a discrete system is exactly analogous to that for a lumped system. The z-transform is to the discrete (difference) system just as the Laplace transform is to the lumped (differential) system.

The specific steps are readily seen in the second-order case. Here

$$x_1(k+1) = a_{11}x_1(k) + a_{12}x_2(k) + u_1(k)$$
$$x_2(k+1) = a_{21}x_1(k) + a_{22}x_2(k) + u_2(k) \tag{5-72}$$

Let the eigenvalues of the $\underset{\sim}{A}$ matrix be distinct. The formula for these eigenvalues is

$$q_1 = \frac{a_{11} + a_{22}}{2} + \sqrt{\frac{(a_{11} + a_{22})^2}{4} - |\underset{\sim}{A}|}$$
$$q_2 = \frac{a_{11} + a_{22}}{2} - \sqrt{\frac{(a_{11} + a_{22})^2}{4} - |\underset{\sim}{A}|} \tag{5-73}$$

For the variation of parameters step we assume

$$x_1(k) = \beta_1(k)\ q_1^k + \beta_2(k)\ q_2^k$$
$$x_2(k) = \frac{q_1 - a_{11}}{a_{12}} \beta_1(k)\ q_1^k + \frac{q_2 - a_{11}}{a_{12}} \beta_2(k)\ q_2^k \tag{5-74}$$

where β_1 and β_2 are to be determined by the method.

Proceeding with the variation of parameters method with completely general coefficients obscures the basic steps with very messy formulas. Thus we shall use specific numbers for the remainder of this example. Let

$$\underset{\sim}{A} = \begin{bmatrix} \frac{1}{2} & \frac{1}{3} \\ \frac{1}{3} & \frac{1}{2} \end{bmatrix} \tag{5-75}$$

Then

$$q_1 = \tfrac{5}{6} \qquad q_2 = \tfrac{1}{6} \tag{5-76}$$

With these numbers (5-74) becomes

$$x_1(k) = \beta_1(k)\ (\tfrac{5}{6})^k + \beta_2(k)\ (\tfrac{1}{6})^k$$
$$x_2(k) = \beta_1(k)\ (\tfrac{5}{6})^k - \beta_2(k)\ (\tfrac{1}{6})^k \tag{5-77}$$

Substituting (5-77) in (5-72) with $\underset{\sim}{A}$ matrix (5-75) and regrouping gives

$$(\tfrac{5}{6})^{k+1}\ [\beta_1(k+1) - \beta_1(k)] + (\tfrac{1}{6})^{k+1}\ [\beta_2(k+1) - \beta_2(k)] = u_1(k)$$
$$(\tfrac{5}{6})^{k+1}\ [\beta_1(k+1) - \beta_1(k)] - (\tfrac{1}{6})^{k+1}\ [\beta_2(k+1) - \beta_2(k)] = u_2(k) \tag{5-78}$$

These two equations are linear algebraic equations with $[\beta_1(k+1) - \beta_1(k)]$ and $[\beta_2(k+1) - \beta_2(k)]$ as unknowns. They can be solved by *Cramers rule*. The result is

$$\beta_1(k+1) - \beta_1(k) = \tfrac{3}{5}(\tfrac{5}{6})^{-k}u_1(k) + \tfrac{3}{5}(\tfrac{5}{6})^{-k}u_2(k)$$
$$\beta_2(k+1) - \beta_2(k) = 3(\tfrac{1}{6})^{-k}u_1(k) - 3(\tfrac{1}{6})^{-k}u_2(k) \tag{5-79}$$

These two equations can be summed to give $\beta_1(k)$ and $\beta_2(k)$ via the formulas (5-62). When (5-62a) is used, the result is

$$\beta_1(k) = \sum_{n=1}^{\infty} \tfrac{3}{5}(\tfrac{5}{6})^{-(k-n)} [u_1(k-n) + u_2(k-n)]$$
$$\beta_2(k) = \sum_{n=1}^{\infty} 3(\tfrac{1}{6})^{-(k-n)} [u_1(k-n) - u_2(k-n)] \tag{5-80}$$

The final solution is found by substituting (5-80) into (5-77). The result is

$$x_1(k) = \sum_{n=1}^{\infty} \{[\tfrac{3}{5}(\tfrac{5}{6})^n + 3(\tfrac{1}{6})^n]\, u_1(k-n) + [\tfrac{3}{5}(\tfrac{5}{6})^n - 3(\tfrac{1}{6})^n]\, u_2(k-n)\}$$
$$x_2(k) = \sum_{n=1}^{\infty} \{[\tfrac{3}{5}(\tfrac{5}{6})^n - 3(\tfrac{1}{6})^n]\, u_1(k-n) + [\tfrac{3}{5}(\tfrac{5}{6})^n + 3(\tfrac{1}{6})^n]\, u_2(k-n)\} \tag{5-81}$$

Z-transforms of matrix difference equations. A shortcut to arrive at the solution (5-81) can be obtained by Z-transformation of the matrix difference equation (5-72) and solution in the Z_1-domain. Transforming gives

$$zX_1(z) = a_{11}\, X_1(z) + a_{12}\, X_2(z) + U_1(z)$$
$$zX_2(z) = a_{21}\, X_1(z) + a_{22}\, X_2(z) + U_2(z) \tag{5-82}$$

Solving gives

$$X_1(z) = \frac{(z-a_{22})}{(z-q_1)(z-q_2)}\, U_1(z) + \frac{a_{12}}{(z-q_1)(z-q_2)}\, U_2(z)$$
$$X_2(z) = \frac{a_{21}}{(z-q_1)(z-q_2)}\, U_1(z) + \frac{z-a_{11}}{(z-q_1)(z-q_2)}\, U_2(z) \tag{5-83}$$

where q_1 and q_2 are the eigenvalues given by (5-73).

To compare (5-83) with (5-81) let us find the discrete time function that corresponds to $[(z-a_{22})/(z-q_1)(z-q_2)]\, U_1(z)$. The result should be identical with the dependence of $x_1(k)$ on $u_1(k)$ in (5-81). The first step in the comparison is to note that if

$$f(k) = \begin{cases} 0 & \text{for } k < 0 \\ q^k & \text{for } k \geqslant 0 \end{cases}$$
$$F_1(z) = \frac{z}{z-q} \tag{5-84}$$

By the shifting theorem (5-32) the function whose Z_1-transform is $1/(z-q)$ is

$$Z_1^{-1}\left(\frac{1}{z-q}\right) = \begin{cases} 0 & \text{for } k < 1 \\ q^{k-1} & \text{for } k \geq 1 \end{cases} \tag{5-85}$$

Thus

$$Z_1^{-1}\left[\frac{z-\frac{1}{2}}{(z-\frac{5}{6})(z-\frac{1}{6})}\right] = \begin{cases} 0 & \text{for } k < 1 \\ \frac{1}{2}(\frac{5}{6})^{k-1} + \frac{1}{2}(\frac{1}{6})^{k-1} & \text{for } k \geq 1 \end{cases} \tag{5-86}$$

The resulting time domain term is

$$\begin{aligned} Z_1^{-1}\left[\frac{z-\frac{1}{2}}{(z-\frac{5}{6})(z-\frac{1}{6})} U_1(z)\right] &= \sum_{n=1}^{\infty} \tfrac{1}{2}[(\tfrac{5}{6})^{n-1} + (\tfrac{1}{6})^{n-1}]\, U_1(k-n) \\ &= \sum_{n=1}^{\infty} [\tfrac{3}{5}(\tfrac{5}{6})^{n} + 3(\tfrac{1}{6})^{n}]\, U_1(k-n) \end{aligned} \tag{5-87}$$

This checks Equation (5-81).

PROBLEMS

5-1 **a.** Carefully derive Equation (5-43).

b. Show that if $h(t) = \begin{cases} e^{p_1 t} - e^{p_2 t} & \text{for } t \geq 0 \\ 0 & \text{for } t \leq 0 \end{cases}$ then the extra term in the error (5-44) drops out.

5-2 For a pulse width modulator, show the error in the approximate analysis when

$$h(t) = \begin{cases} te^{pt} & \text{for } t \geq 0 \\ 0 & \text{for } t \leq 0 \end{cases}$$

5-3 Fill in Table 5-1 for pulse position modulation. For those blanks that cannot be filled by a convenient formula, explain why.

5-4 Find the inverse Z-transform of (5-54).

5-5 Consider the system of Figure P5-5 with $h(t)$ and $v(t)$ given by (5-47) and (5-48), respectively, and the modulators sample and hold with $T = 1$.

a. Find the Z-transform of the output and compare it with (5-54).

b. Find $x(t)$ and compare with the results of Problem 5-4.

5-6 Apply the approximate formulas to the example of Figure 5-2 and of Problem 5-5. Compare the responses to the results of Problems 5-4 and 5-5.

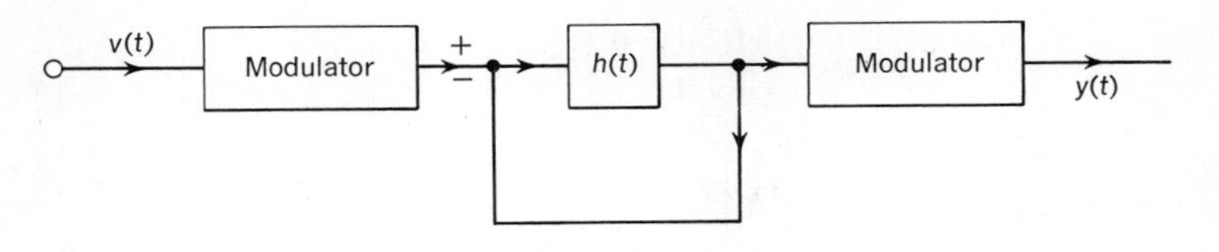

Figure P5-5

5-7 Consider Figure 5-2 with pulse width modulators. Let $h(t)$ be given by (5-47). Estimate the error that will result when the approximate analysis is used. Choose an appropriate scaling factor a when the sampling period is one, and the input is given by (5-48) with a bias so that $v(t)$ is never negative.

5-8 Consider a discrete approximation to a first-order differential system. Show how the discrete step width effects the approximation to the response. Specifically start with the system $\dot{x} = ax + u$. Approximate $\dot{x}(kT)$ by $\dfrac{y((k+1)T) - y(kT)}{T}$. The result is the difference equation

$$y((k+1)T) = (aT+1)\, y(kT) + Tu(kT)$$

Compare $y(t)$, the solution to the difference equation, with $x(t)$, the solution to the differential equation for different values of the period T and the natural frequency $-a$.

5-9 Find the total response of the discrete time system characterized by the equations

$$\mathbf{x}(k+1) = \begin{bmatrix} -\frac{1}{6} & -\frac{1}{6} \\ \frac{1}{3} & -\frac{2}{3} \end{bmatrix} \mathbf{x}(k) + \begin{bmatrix} 1 \\ 0 \end{bmatrix} v(k)$$

$$y(k) = [1 \quad 2]\; x(k) + v(k) + \tfrac{1}{2}\, v(k+1)$$

The initial state is $\mathbf{x}(0) = \begin{bmatrix} 1 \\ 0 \end{bmatrix}$

The input is $v_k = \begin{cases} k & \text{for } 0 \leqslant k \leqslant 3 \\ 6-k & \text{for } 3 \leqslant k \leqslant 6 \\ 0 & \text{for } k > 6 \end{cases}$

5-10 Find the total response of the discrete time system characterized by

$$\mathbf{x}(k+1) = \begin{bmatrix} \frac{1}{2} & \frac{1}{2} & 0 \\ -\frac{1}{2} & \frac{1}{2} & 0 \\ 0 & 0 & \frac{1}{2} \end{bmatrix} \mathbf{x}(k) + \begin{bmatrix} 1 \\ 0 \\ 1 \end{bmatrix} v(k)$$

$$\mathbf{v}(k) = \begin{bmatrix} 1 & 2 & 0 \\ 1 & 1 & 1 \end{bmatrix} \mathbf{x}(k)$$

The initial state is $\mathbf{x}(0) = \begin{bmatrix} 1 \\ 0 \\ 1 \end{bmatrix}$

The input is $v(k) = \cos\ (k/10)$ for $k > 0$

5-11 When the $\underset{\sim}{A}$ matrix for a discrete time system has complex eigenvalues, these must occur in conjugate pairs. Furthermore, the corresponding coefficients in the response functions must be conjugated so that the response is real for real inputs. Write a closed form *real* expression for

$$K\ q^{kT} + K^*\ (q^*)^{kT}$$

5-12 For a single-input–single-output discrete time system, the output $y(k)$ has the form

$$y(k) = \sum_{i=1}^{n} \left[\sum_{j=-\infty}^{k} K_i\ q_i^{k-j}\ v(j) \right] + Dv(k) + Ev(k+1)$$

where the q_i are eigenvalues of the $\underset{\sim}{A}$ matrix and $v(j)$ is the input. The sum on j is taken from $-\infty$ to account for all inputs. The constants K_i correspond to the constants in the response function for lumped, linear systems. Thus, block diagrams for discrete time systems can be constructed by analogy to the lumped, linear systems.

For the discrete time block diagram of Figure P5-12 set up normal form equations in the form

$$\mathbf{x}(k+1) = \underset{\sim}{A}\ \mathbf{x}(k) + \underset{\sim}{B}\ u(k)$$
$$y(k) = \underset{\sim}{C}\ \mathbf{x}(k) + D\ u(k)$$

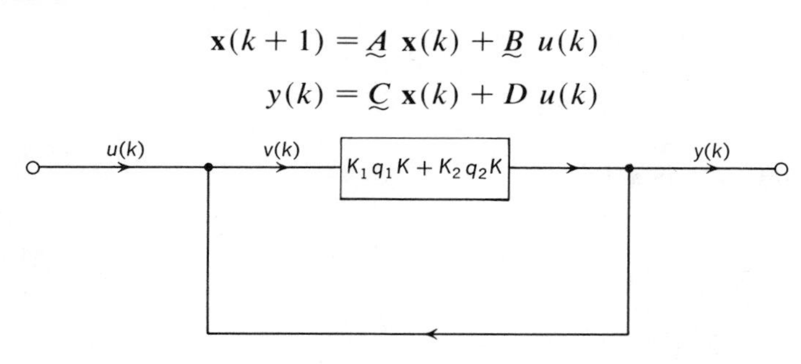

Figure P5-12

PART II

Some Classifications for Lumped, Linear, Time-Invariant Systems

CHAPTER

6

Systems with Positive-Semidefinite Energy Functions

6-1 Preliminaries

In Part I, the basic emphasis was quantitative analysis; that is, given a system diagram including the input signals, find the response. In this chapter and Chapter 8 the basic emphasis is qualitative analysis; that is, given a system whose equations have certain properties, find other important properties of the system. Passivity and stability are two of the properties of interest. These two properties, along with others, will be defined as needed.

In this chapter we consider systems in which power and energy are significant quantities. Electrical and mechanical systems both fall in this category. Each element in such a system either generates energy, stores energy, or dissipates energy. In such a system Kirchhoff's laws or D'Alembert's principal are the basis for forming equilibrium equations from an elemental diagram. For each element there are two signals. The product of the values of the signals is the power or energy associated with that element. When the product is power, the signals are current and voltage for electrical systems, and force and velocity for mechanical systems. When the product is energy, then the signals are charge-voltage, flux linkage-current, momentum-velocity, or position-force.

Power and energy are well-defined for the complete system as well as for each element. The inputs and outputs of these systems are taken at ports. Each port has the two signals whose product is the power or energy associated with it. To be specific in this chapter the input and output signals will be currents and voltages or forces and velocities. Furthermore, at each port one signal will be considered as an input and the other an output. The sign for the signals will be chosen so that the product of the two signals at a port is the power delivered to the system at that port. In conventional electrical network terminology this means that our systems will be characterized on an admittance, impedance, or hybrid parameter basis.

More specifically we consider the class of linear, time-invariant systems that can be described as follows:

$$\begin{aligned} \dot{\mathbf{x}} &= \underset{\sim}{A}\,\mathbf{x} + \underset{\sim}{B}\,\mathbf{v} \\ \mathbf{y} &= \underset{\sim}{C}\,\mathbf{x} + \underset{\sim}{D}\,\mathbf{v} + \underset{\sim}{E}\,\dot{v} \end{aligned} \tag{6-1}$$

where $\mathbf{x}$ is an n-vector

$\underset{\sim}{A}$ is an $n \times n$ matrix
$\underset{\sim}{B}$ is an $n \times p$ matrix
$\mathbf{v}$ is a p-vector
$\mathbf{y}$ is a p-vector
$\underset{\sim}{C}$ is a $p \times n$ matrix
$\underset{\sim}{D}$ and $\underset{\sim}{E}$ are $p \times p$ matrixes.

The power delivered to the system is the sum of products of the respective input and output quantities at each port. Thus

$$P = \mathbf{v}^T\,\mathbf{y} \tag{6-2}$$

where the superscript T means transpose. This description covers all the electric circuits and their mechanical analogs that can be put into state variable form by the methods of Chapter 2.

By proper choice of state variables, as discussed below, the energy stored in the system is

$$W = \tfrac{1}{2}\,\mathbf{x}^T\,\mathbf{x} \tag{6-3}$$

The power dissipated by a system is the difference between the total power delivered and that which goes into storage. The power into storage is the rate of change of energy storage. Thus the dissipated power is

$$P_d = P - \dot{W} \tag{6-4}$$

The dissipated power given by (6-4) must be equal to the total power dissipated in the individual elements if the system is consistent with the laws of physics. For electric networks this consistency is known as Tellegen's theorem (see Reference 26).

In this chapter we consider systems for which the energy storage and the dissipated power are never negative. We shall see that such systems are passive and stable and that the transfer function matrix has special properties. In order to build up the techniques gradually we begin with resistive systems, then we consider lossless systems, and finally we consider the general case of (6-1).

6-2 Resistive Systems

A resistive system is characterized by a zero-dimensional state vector and p-dimensional $\mathbf{v}$ and $\mathbf{y}$ vectors. The input-output relations are the algebraic equations

$$\mathbf{y} = \underset{\sim}{D}\, \mathbf{v}$$

The power delivered to the system at the ports is

$$P = \mathbf{v}^T \mathbf{y} = \mathbf{v}^T \underset{\sim}{D}\, \mathbf{v} = \sum_{i=1}^{p} \sum_{j=1}^{p} d_{ij}\, v_i v_j \tag{6-5}$$

The vector $\mathbf{v}$ defines a point in a p-dimensional space. For each such point the power P has a value. Thus, the power is a function of p-variables; it is a surface in a $(p+1)$-dimensional space. Since each term in the sum on the right of (6-5) involves a product of exactly two components (one component times itself included) of $\mathbf{v}$, P is called a quadratic form.

From physical arguments it is obvious that for a circuit of positive resistors, P is positive for all excitations except $\mathbf{v} = \mathbf{0}$ when $P = 0$. Quadratic forms of this type have been studied extensively. Such functions are called positive definite (PD) quadratic forms. Any linear algebra book will have a discussion of quadratic forms and criteria that the matrix $\underset{\sim}{D}$ in (6-5) must satisfy for the associated form to be PD. A resistive system is passive so long as the power is non-negative. A quadratic form of this type; namely, one that may be zero even though $\mathbf{v}$ is nonzero, is called positive semidefinite (PSD).

Tests for positive definiteness. The basic idea behind tests for positive definiteness or semidefiniteness of matrixes is the fact that we can always make a change of variables in the function (6-5). If the change results in a quadratic form with a diagonal matrix, then the resultant is readily checked. In matrix notation a change of variables from $\mathbf{v}$ to $\mathbf{z}$ is written

$$\mathbf{v} = \underset{\sim}{K}\, \mathbf{z} \tag{6-6}$$

If $\underset{\sim}{K}$ is nonsingular (its determinant is nonzero), then the equations represented by (6-6) can be solved

$$\mathbf{z} = \underset{\sim}{K}^{-1} \mathbf{v} \tag{6-7}$$

Thus as $\mathbf{v}$ goes through all real vector values, $\mathbf{z}$ also goes through all values.

Substituting (6-6) into (6-5) gives

$$P = \mathbf{z}^T \underset{\sim}{F} \mathbf{z} \tag{6-8}$$

where

$$\underset{\sim}{F} = \underset{\sim}{K}^T \underset{\sim}{D} \underset{\sim}{K}$$

Now (6-8) takes on the same set of values as (6-5) as $\mathbf{z}$ or $\mathbf{v}$ take on all values. Thus the *PD*ness or *PSD*ness of F and D are the same. If we can find a $\underset{\sim}{K}$ so that $\underset{\sim}{F}$ is diagonal, the test is trivial on $\underset{\sim}{F}$ because

$$P = \sum_{i=1}^{p} F_{ii}\, z_i^2 \tag{6-9}$$

A thorough discussion of quadratic forms as they apply to system theory problems is given in Chapter 4 of Reference 12. An example illustrates some of the important considerations.

Consider the transistor circuit of Figure 6-1. The problem is to find

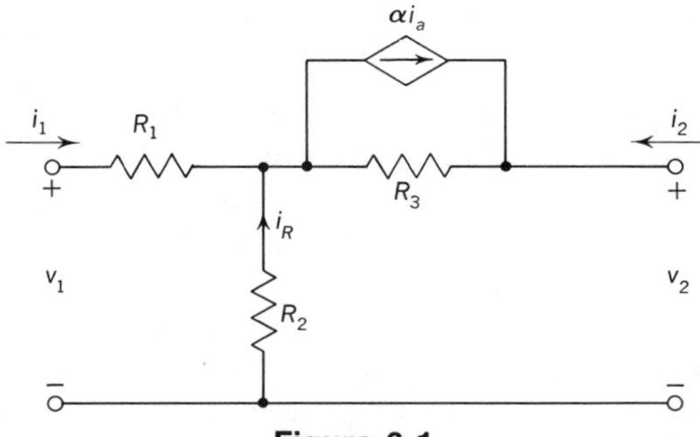

Figure 6-1

relations among the parameters R_1, R_2, R_3, and α, for which the circuit is, or is not passive.[1] For this circuit the impedance matrix is

$$\underset{\sim}{Z} = \begin{bmatrix} R_1 + R_2 & R_2 \\ R_2 - \alpha R_3 & R_2 + (1 - \alpha) R_3 \end{bmatrix}$$

The power delivered to the ports is

[1] For a resistive system $P \geqslant 0$ for all possible inputs is a criterion for passivity. Passivity is defined for the general system of (6-1) in Section 6-4 below.

$$P = \mathbf{i}^T\mathbf{v} = \mathbf{i}^T \underset{\sim}{Z}\, \mathbf{i}$$
$$= (R_1 + R_2)\, i_1^2 + (R_2 + R_2 - \alpha R_3)\, i_1 i_2 + (R_2 + (1 - \alpha)\, R_3)\, i_2^2 \qquad \textbf{(6-10)}$$

The procedure for finding the transformation matrix $\underset{\sim}{K}$ that generates the diagonal matrix $\underset{\sim}{F}$ for the form (6-9) is most easily carried out if the original quadratic form is generated from a symmetric matrix. Although (6-10) was generated by an asymmetric matrix, the same form could have been generated by the symmetric matrix

$$\begin{bmatrix} R_1 + R_2 & R_2 - \dfrac{\alpha}{2} R_3 \\ R_2 - \dfrac{\alpha}{2} R_3 & R_2 + (1 - \alpha)\, R_3 \end{bmatrix}$$

There are infinitely many transformation matrixes $\underset{\sim}{K}$ that will generate a diagonal $\underset{\sim}{F}$ matrix. If the original matrix $\underset{\sim}{D}$ is symmetric then any matrix $\underset{\sim}{K}^T \underset{\sim}{D}\, \underset{\sim}{K}$ is symmetric. In the 2×2 case, the off-diagonal term in $\underset{\sim}{K}^T \underset{\sim}{D}\, \underset{\sim}{K}$ is

$$k_{11} k_{12} d_{11} + (k_{12} k_{21} + k_{22} k_{11})\, d_{12} + k_{22} k_{21} d_{22}$$

To generate the diagonal matrix $\underset{\sim}{F}$ the k_{ij} must be chosen so that this term is zero and so that the determinant of $\underset{\sim}{K}$, $(k_{11} k_{22} - k_{12} k_{21})$, is nonzero. Any such selection will do. An easy selection is $k_{12} = 0, k_{11} = 1$. Then

$$k_{21} = -\frac{d_{12}}{d_{22}} \qquad \textbf{(6-11)}$$

Any nonzero k_{22} will do, so let $k_{22} = 1$.

For the example in question

$$F = \begin{bmatrix} 1 & -\dfrac{R_2 - (\alpha/2) R_3}{R_2 + (1 - \alpha)\, R_3} \\ 0 & 1 \end{bmatrix} \begin{bmatrix} R_1 + R_2 & R_2 - \dfrac{\alpha}{2} R_3 \\ R_2 - \dfrac{\alpha}{2} R_3 & R_2 + (1 - \alpha)\, R_3 \end{bmatrix}$$
$$\begin{bmatrix} 1 & 0 \\ -\dfrac{R_2 - (\alpha/2)\, R_3}{R_2 + (1 - \alpha)\, R_3} & 1 \end{bmatrix}$$

$$= \begin{bmatrix} R_1 + R_2 - \dfrac{(R_2 - (\alpha/2)\, R_3)^2}{R_2 + (1 - \alpha)\, R_3} & 0 \\ 0 & R_2 + (1 - \alpha)\, R_3 \end{bmatrix}$$

If the circuit is to be passive, both of the terms in $\underset{\sim}{F}$ must be positive or zero. We should note that this means d_{22} and the determinant of $\underset{\sim}{D}$ must be positive or zero. The generalization of these conditions to a test on a sequence of determinants of degree one, two, up to p is called *Sylvester's theorem*. It applies to *PD* and *PSD* forms in the 2×2 case, but only to *PD* forms in higher-order situations.

To put reasonable numbers in the problem, let $R_1 = 1, R_2 = 1$, $R_3 = 100$. Then for passivity $f_{22} \geqslant 0$ requires $\alpha \leqslant 1.01$; $f_{11} \geqslant 0$ gives

$$2 + 200\,(1 - \alpha) - 1 + 100\,\alpha - \frac{10^4}{4}\,\alpha^2 \geqslant 0$$

For positive α this requires $\alpha < 0.264$. Thus, a transistor with the resistance ratios $R_e = R_b = R_c/100$ cannot be a power amplifier under any source-load conditions if $\alpha < 0.264$.

The above discussion should be adequate for the physical significance of quadratic forms in resistive circuits. Some of these ideas will apply equally to energy storage in more general systems. The mathematics of quadratic forms is much more extensive. Since there are many excellent treatments of the mathematical aspects, we shall not pursue the matter further (see Reference 12 for a readable treatment of the subject for those with engineering backgrounds).

6-3 Lossless Systems

For a lossless system the power delivered to the system is stored as energy. Such a system has two types of energy storage—electric and magnetic for electrical systems or kinetic and potential for mechanical systems. In terms of system parameters and variables the stored energies for lumped, linear, time-invariant systems are

$$V = \text{Potential energy} = \tfrac{1}{2}\,K\,x^2 = \tfrac{1}{2}\,\frac{f^2}{K} \tag{6-12}$$

where f is the force; x is the displacement; K is the spring constant or potential field constant.

$$T = \text{Kinetic energy} = \tfrac{1}{2}\,v\,p = \tfrac{1}{2}\,M\,v^2 = \tfrac{1}{2}\,\frac{p^2}{K} \tag{6-13}$$

where v is the velocity; p is the momentum; M is the mass.

$$W_C = \text{Electric energy} = \tfrac{1}{2}\,q\,v = \tfrac{1}{2}\,C\,v^2 = \tfrac{1}{2}\,\frac{q^2}{C} \tag{6-14}$$

where q is the charge; v is the voltage; C is the capacitance.

$$W_L = \text{Magnetic energy} = \tfrac{1}{2}\,\lambda\, i = \tfrac{1}{2}\,L\, i^2 = \tfrac{1}{2}\,\frac{\lambda^2}{L} \tag{6-15}$$

where λ is the flux linkage; i is the current; L is the inductance.

For lossless systems, the state variable formulation is exactly the Hamiltonian formulation of classical mechanics. (Reference 23, Chapter 4, Section 2, is a brief, elementary discussion of Hamilton's equations.) Most of the interesting problems of classical mechanics occur when the system is nonlinear. Then Hamilton's equations are a way to set up normal form equations. In the lumped, linear, time-invariant case the energy ideas are quite simple for lossless systems. For this reason let us go through several examples and see how energy ideas apply before attacking the general case of Equation (6-1).

A mechanical example. As a first example let us look at a simple mechanics problem—the pendulum of Figure 6-2. By using Newton's

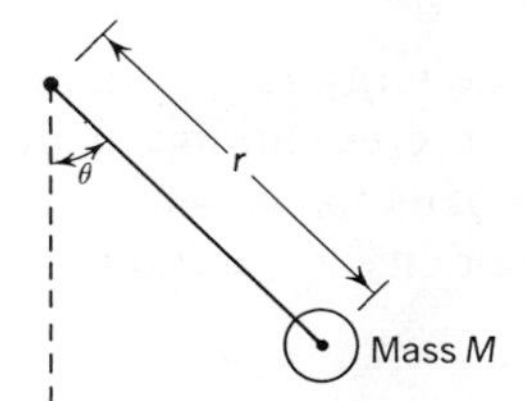

Figure 6-2

law we can easily set up the equations of motion. The motion is in the radial direction. The force in that direction is

$$F = M\,g\,\sin\theta$$

where g is the acceleration of gravity. When θ is positive, this force is in the negative direction. Thus Newton's law gives

$$M\,g\,\sin\theta = -Mr\,\dot{\omega}$$
$$\omega = \dot{\theta}$$

In normal form the equations are

$$\begin{aligned} \dot{\theta} &= \omega \\ \dot{\omega} &= -\frac{g}{r}\sin\theta \end{aligned} \tag{6-16}$$

For this system the kinetic energy is

$$T = \tfrac{1}{2}\,M\,(r\omega)^2 \tag{6-17}$$

The potential energy must be established with respect to some reference potential. If this reference is chosen at the point $\theta = 0$ then

$$V = Mgr\ (1 - \cos\theta) \tag{6-18}$$

This last form is not the same as that of a lumped, linear, system (6-12). In fact, with a different choice of reference, V can be positive or negative whereas (6-12) is always positive. Many of the energy ideas apply to systems other than those of (6-1).

To show that energy is conserved we compute the derivative of the total energy. It is

$$\frac{d}{dt}[\tfrac{1}{2}M(r\omega)^2 + Mgr(1 - \cos\theta)] = Mr(r\omega\dot{\omega} + g\sin\theta\ \dot{\theta}) \tag{6-19}$$

Substituting for ω and $\dot{\omega}$ from (6-16) gives

$$r\omega\dot{\omega} + g\sin\theta\ \dot{\theta} = r\dot{\theta}\left(-\frac{g}{r}\sin\theta\right) + g\sin\theta\ \dot{\theta} = 0$$

Thus the rate of change of energy is zero and the system is lossless.

Often for small swings the equations of motion (6-16) are linearized by assuming $\sin\theta = \theta$. Under this assumption the force is $M\ g\ \theta$ and the displacement is $r\ \theta$. The velocity is $r\omega$ and the momentum $Mr\omega$. The total energy is

$$T + V = \tfrac{1}{2}[Mgr\ \theta^2 + Mr^2\ \omega^2]$$

The derivative of the total energy is

$$\dot{T} + \dot{V} = Mr(g\ \theta\ \dot{\theta} + r\omega\ \dot{\omega}) \tag{6-20}$$

The linearized equations of motion from (6-16) are

$$\begin{aligned} \dot{\theta} &= \omega \\ \dot{\omega} &= -\frac{g}{r}\theta \end{aligned} \tag{6-21}$$

Substituting (6-20) into (6-21) shows that the system is still lossless.

Lossless electric circuits. To see the connections between energy functions and the topological formulation of state equations let us consider an $L-C$ circuit excited by current sources across the capacitors and voltage sources in series with the inductors. To keep the notation simple, let us assume there are no capacitance loops or inductance cut sets. We denote the inductor currents by i_j, the capacitor voltages by v_k, the source currents by I_k, and the source voltages by V_j. Let there be J inductors and K capacitors. Then $\mathbf{i}$ is a J-dimensional vector as is $\mathbf{V}$. Both $\mathbf{v}$ and $\mathbf{I}$ are K-dimensional. Now the normal form is

$$\begin{bmatrix} \dot{\mathbf{i}} \\ \dot{\mathbf{v}} \end{bmatrix} = \begin{bmatrix} \underset{\sim}{Q}_{JJ} & \underset{\sim}{A}_1 \\ \underset{\sim}{A}_2 & \underset{\sim}{Q}_{KK} \end{bmatrix} \begin{bmatrix} \mathbf{i} \\ \mathbf{v} \end{bmatrix} + \begin{bmatrix} \dfrac{1}{\underset{\sim}{L}} & \underset{\sim}{Q}_{JK} \\ \underset{\sim}{Q}_{KJ} & \dfrac{1}{\underset{\sim}{C}} \end{bmatrix} \begin{bmatrix} \mathbf{V} \\ \mathbf{I} \end{bmatrix} \tag{6-22}$$

where the $\underset{\sim}{Q}_{xy}$ is a zero matrix with x rows and y columns. The matrixes $1/\underset{\sim}{L}$ and $1/\underset{\sim}{C}$ are $J \times J$ and $K \times K$, respectively, diagonal matrixes, with the appropriate $1/L_j$ or $1/C_k$ elements. The energy stored in the inductors is

$$W_L = \mathbf{i}^T \underset{\sim}{L}\, \mathbf{i} \tag{6-23}$$

Similarly for capacitors

$$W_C = \tfrac{1}{2}\, \mathbf{v}^T \underset{\sim}{C}\, \mathbf{v} \tag{6-24}$$

Here both $\underset{\sim}{C}$ and $\underset{\sim}{L}$ are diagonal matrixes.

Since the circuit is lossless, the power supplied by the sources must be equal to the rate of change of stored energy. Thus

$$\mathbf{i}^T \mathbf{V} + \mathbf{v}^T \mathbf{I} = \dot{W}_L + \dot{W}_C \tag{6-25}$$

In the special case under consideration where $\underset{\sim}{L}$ and $\underset{\sim}{C}$ in (6-23) and (6-24) are diagonal.

$$W_L = \tfrac{1}{2} \sum_{i=1}^{J} L_{ii}\, i_i^2 \text{ and } W_C = \tfrac{1}{2} \sum_{i=1}^{K} C_{ii}\, v_i^2$$

Then

$$\left.\begin{aligned} \dot{W}_L &= \sum_{i=1}^{J} L_{ii}\, i_i\, \dot{i}_i = \mathbf{i}^T \underset{\sim}{L}\, \dot{\mathbf{i}} \\ &\text{and} \\ \dot{W}_c &= \mathbf{v}^T \underset{\sim}{C}\, \dot{\mathbf{v}} \end{aligned}\right\} \tag{6-26}$$

With (6-26) and (6-22) the right side of (6-25) can be written as

$$\begin{aligned} \dot{W}_L + \dot{W}_c &= [\mathbf{i}^T \quad \mathbf{v}^T] \begin{bmatrix} \underset{\sim}{L} & \underset{\sim}{Q}_{JK} \\ \underset{\sim}{Q}_{KJ} & \underset{\sim}{C} \end{bmatrix} \begin{bmatrix} \dot{\mathbf{i}} \\ \dot{\mathbf{v}} \end{bmatrix} \\ &= [\mathbf{i}^T \quad \mathbf{v}^T] \begin{bmatrix} \underset{\sim}{L} & \underset{\sim}{Q}_{JK} \\ \underset{\sim}{Q}_{KJ} & \underset{\sim}{C} \end{bmatrix} \begin{bmatrix} \underset{\sim}{Q}_{JJ} & \underset{\sim}{A}_1 \\ \underset{\sim}{A}_2 & \underset{\sim}{Q}_{KK} \end{bmatrix} \begin{bmatrix} \mathbf{i} \\ \mathbf{v} \end{bmatrix} \\ &\quad + [\mathbf{i}^T \quad \mathbf{v}^T] \begin{bmatrix} \underset{\sim}{L} & \underset{\sim}{Q}_{JK} \\ \underset{\sim}{Q}_{KJ} & \underset{\sim}{C} \end{bmatrix} \begin{bmatrix} \dfrac{1}{\underset{\sim}{L}} & \underset{\sim}{Q}_{JK} \\ \underset{\sim}{Q}_{KJ} & \dfrac{1}{\underset{\sim}{C}} \end{bmatrix} \begin{bmatrix} \mathbf{V} \\ \mathbf{I} \end{bmatrix} \end{aligned} \tag{6-27}$$

Since $\underset{\sim}{L}$ and $1/\underset{\sim}{L}$; $\underset{\sim}{C}$ and $1/\underset{\sim}{C}$ are diagonal matrixes with reciprocal elements, they are inverses, respectively. Thus, the last expression in (6-27) is the left side of (6-25). If the physical justification of (6-25) is correct, then we must be able to show that

$$[\mathbf{i}^T \quad \mathbf{v}^T] \begin{bmatrix} \underset{\sim}{L} & \underset{\sim}{O}_{JK} \\ \underset{\sim}{O}_{KJ} & \underset{\sim}{C} \end{bmatrix} \begin{bmatrix} \underset{\sim}{O}_{JJ} & \underset{\sim}{A}_1 \\ \underset{\sim}{A}_2 & \underset{\sim}{O}_{KK} \end{bmatrix} \begin{bmatrix} \mathbf{i} \\ \mathbf{v} \end{bmatrix} = 0 \tag{6-28}$$

The truth of (6-28) is a consequence of the topology of the $L-C$ network. If there are J inductors and K capacitors, then the first j rows of

$$\begin{bmatrix} \underset{\sim}{O} & \underset{\sim}{A}_1 \\ \underset{\sim}{A}_2 & \underset{\sim}{O} \end{bmatrix} \begin{bmatrix} \mathbf{i} \\ \mathbf{v} \end{bmatrix}$$

have the form $(1/L_j)\ \Sigma$ [tree branch voltages across the jth link] $j = 1, 2, \cdots J$. Similarly, the last K rows have the form $(1/C_k)\ \Sigma$ (link currents through the kth branch) $k = 1, 2, \cdots, K$. Multiplying by

$$\begin{bmatrix} \underset{\sim}{L} & \underset{\sim}{O} \\ 0 & \underset{\sim}{C} \end{bmatrix}$$

cancels the $1/C_k$ and $1/L_j$ multiplying each row, respectively. The final quadratic form has the form

$$\Sigma\ \{\text{link currents}\ [\Sigma\ (\text{tree branch voltages across})] + \text{tree branch voltages}\ [\Sigma\ (\text{link currents through})]\}$$

This sum is zero because when a tree branch voltage is across a link with a positive orientation, the corresponding link current goes through the tree branch in the negative direction.

When a circuit has capacitor loops, inductor cut sets, or coupled coils, equations written directly from a tree will not be in normal form. When the derivative terms are put on the left and the other terms on the right, there will be some equations with derivatives of two or more of the variables. Except in the case of transformers with unity coupling, the nondiagonal $\underset{\sim}{L}$ or $\underset{\sim}{C}$ matrix can be diagonalized by a change of variables like (6-6), which led to the diagonal quadratic form (6-9). If such a change of variables is made and the diagonal matrix that results used for the

$$\begin{bmatrix} \underset{\sim}{L} & \underset{\sim}{O} \\ \underset{\sim}{O} & \underset{\sim}{C} \end{bmatrix}$$

matrix above, the state variable equations are still related to the stored energy in the same manner. The notation can be somewhat simplified if, instead of formulating the equations with currents and voltages as state variables, the flux linkages and charges are used. Hamilton's equations are usually written in this latter form.

Circuits with capacitance loops or inductance cut sets. The special problems that arise when the L or C matrixes are not diagonal are best illustrated by an example. Consider the transformer with series tuned primary and shunt tuned secondary shown in Figure 6-3. Assume the

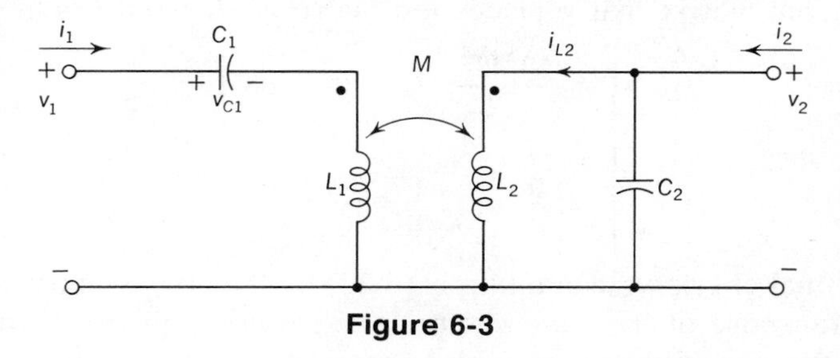

Figure 6-3

left port is excited by a current source, and the right port by a voltage source. The equilibrium equations are

$$\begin{aligned} L_1\,\dot{i}_1 + M\,\dot{i}_{L_2} &= -v_{C_1} + v_1 \\ M\dot{i}_1 + L_2\,\dot{i}_{L_2} &= v_2 \\ C_1\,\dot{v}_{C_1} &= i_1 \\ C_2\,\dot{v}_2 &= -i_{L2} + i_2 \end{aligned} \tag{6-29}$$

The energy stored in the transformer is

$$W_L = \tfrac{1}{2}\,L_1\,i_1^2 + M i_1 i_{L2} + \tfrac{1}{2}\,L_2\,i_{L2}^2 \tag{6-30}$$

With the change of variables like (6-6), the stored magnetic energy can be made into the quadratic form of a diagonal matrix. Using the transformation matrix of (6-11)

$$\begin{bmatrix} i_1 \\ i_{L_2} \end{bmatrix} = \begin{bmatrix} 1 & 0 \\ -\dfrac{M}{L_2} & 1 \end{bmatrix} \begin{bmatrix} j_1 \\ j_2 \end{bmatrix} \tag{6-31}$$

In terms of j_1 and j_2, the stored energy (6-30) becomes

$$W_L = \tfrac{1}{2}\left[L_1 - \frac{M^2}{L_2}\right] j_1^2 + \tfrac{1}{2}\,L_2\,j_2^2 \tag{6-32}$$

The equations (6-29) can be put in normal form with variables j_1, j_2, v_{C_1}, v_2 by substituting (6-31) for i_1 and i_{L_2}. The result is

$$\dot{j}_1 = \frac{L_2}{L_1 L_2 - M^2}\left[-v_{C_1} - \frac{M}{L_2}\,v_2 + v_1\right]$$

$$\dot{j}_2 = \frac{1}{L_2}\,v_2$$

$$\dot{v}_{C_1} = \frac{1}{C_1}\,j_1$$

$$\dot{v}_2 = \frac{M}{L_2 C_2} j_1 - \frac{1}{C_2} j_2 + \frac{1}{C_2} i_2 \tag{6-33}$$

The diagonal matrix that replaces the matrix of L's and C's in (6-27) is

$$\begin{bmatrix} L_1 - \dfrac{M^2}{L_2} & 0 & 0 & 0 \\ 0 & L_2 & 0 & 0 \\ 0 & 0 & C_1 & 0 \\ 0 & 0 & 0 & C_2 \end{bmatrix}$$

As a final check we can multiply (6-33) by the above matrix and then by the transpose of the state vector. The resulting quadratic form, like (6-27) is the rate of change of stored energy. Since the system is lossless, the result of these operations should be equal to the power input. It is

$$\begin{aligned} j_1 \left(L_1 - \frac{M^2}{L_2}\right) \dot{j}_1 + L_2 \, j_2 \, \dot{j}_2 + v_{C_1} \, C_1 \, \dot{v}_{C_1} + v_2 \, C_2 \, \dot{v}_2 & \\ = -j_1 \, v_{C_1} - \frac{M}{L_2} j_1 \, v_2 + j_1 \, v_1 + j_2 \, v_2 + v_{C_1} j_1 & \\ + \frac{M}{L_2} v_2 \, j_1 - v_2 \, j_2 + v_2 \, i_2 & \\ = j_1 \, v_1 + v_2 \, i_2 & \end{aligned}$$

From (6-31) we see that $j_1 = i_1$. Thus $j_1 \, v_1 + v_2 \, i_2$ is the power input at the 2-ports of Figure 6-3.

The above example shows, at least for the two coupled energy storage elements case, that state equations can be written in such a way that the stored energy is a diagonal quadratic form in the state variables. In fact, except for some very special cases as indicated in the next section, inductor cut sets, capacitor loops, differential gears, and other similar coupled energy storage elements can be handled similarly.

6-4 The General System of Equation (6-1)

For the lumped, linear, time-invariant system that can be modeled (possibly on an analog basis) by R, L, C, M, and controlled sources α, μ, r_m, g_m, the model is a network of branches satisfying Kirchhoff's laws and except in very special cases admitting a formulation in the form (6-1). Here the input-output relation is on a port basis so that the power is $\mathbf{v}^T\mathbf{y}$. Let the state variables be chosen so that the total stored energy is $\frac{1}{2}\mathbf{x}^T\mathbf{x}$. Such a choice means that the state variable associated with an inductor is $\sqrt{L}\, i$ and with a capacitor is $\sqrt{C}\, v$. When there are capacitor loops, inductor cut sets, or mutual inductors, the situation must be handled as discussed above and by two more examples at the end of this section. We assume there are no points where negative energy is stored. Thus the storage elements must all be positive.

The power and energy relations are

$$\dot{W}_L + \dot{W}_c = \mathbf{x}^T\,\dot{\mathbf{x}} = \mathbf{x}^T\,\underset{\sim}{A}\,\mathbf{x} + \mathbf{x}^T\,\underset{\sim}{B}\,\mathbf{v} \tag{6-34}$$

$$P = \mathbf{v}^T\,\mathbf{y} = \mathbf{v}^T\,\underset{\sim}{C}\,\mathbf{x} + \mathbf{v}^T\,\underset{\sim}{D}\,\mathbf{v} + \mathbf{v}^T\,E\,\dot{\mathbf{v}} \tag{6-35}$$

By definition the dissipated power P_d is the difference between the input power P and the power that goes into storage. Thus

$$P_d = P - (\dot{W}_L + \dot{W}_c) \tag{6-36}$$

Since the total power delivered to a circuit at its ports is the sum of the power delivered to the individual branches, the dissipative power can also be computed as the total power delivered to resistors and controlled sources. This last computation is the power delivered to a resistive network (as discussed in Section 2-2), formed from the original network by replacing all storage elements by ports. The procedure is illustrated by the examples below.

When the energy functions W_L and W_c and the dissipative power P_d are all positive semidefinite, then the circuit is passive, and the transfer function matrix is positive real. Both passivity and positive reality must be defined before these facts can be checked.

Passivity for *PSD* systems. A system is said to be *passive* if for all real excitations $\mathbf{v}(t)$ the net energy delivered to the system from minus infinity to the present time is non-negative. In symbols we define a system as passive if

$$\int_{-\infty}^{t} P(\tau)\; d\tau \geq 0 \qquad \text{for all excitations } \mathbf{v}(t) \tag{6-37}$$

Solving (6-36) for P and substituting into (6-37) shows that if the energy storage is zero at minus infinity, then the passivity criterion becomes

$$W_L + W_c + \int_{-\infty}^{t} P_d(\tau)\; d\tau \geq 0 \tag{6-38}$$

for all excitations $\mathbf{v}(t)$. With the assumption stated above that capacitors and inductors are positive, and thus no negative energy storage is allowed, the passivity or nonpassivity of the system hinges on P_d. If all resistors are positive, controlled sources can still make the system nonpassive. The presence of a controlled source is not a sufficient condition for nonpassivity as shown by the example in Section 6-2 above.

If P_d is positive semidefinite, then the system is surely passive. Furthermore, if P_d is *PSD* then the transfer function matrix has special properties. The transfer function matrix $H(s)$ can be generated from the normal form equations (6-1) by assuming a real time function input of the form

$$\mathbf{v}(t) = \mathbf{V}\, e^{st} + \mathbf{V}^*\, e^{s^*t} \tag{6-39}$$

where the asterisk means complex conjugate. All other quantities have similar functional forms. That is,

$$\begin{aligned} \mathbf{x}(t) &= \mathbf{X}\, e^{st} + \mathbf{X}^*\, e^{s^*t} \\ \mathbf{y}(t) &= \mathbf{Y}\, e^{st} + \mathbf{Y}^*\, e^{s^*t} \end{aligned} \tag{6-40}$$

The vectors of complex numbers **V** and **Y** are related for each s by

$$\mathbf{Y} = \underset{\sim}{H}(s)\, \mathbf{V} \tag{6-41}$$

where $\underset{\sim}{H}(s)$ is the transfer function matrix.

With the state variable chosen as discussed at the beginning of this section so that $\frac{1}{2}\,\mathbf{x}^T\,\mathbf{x}$ is the total stored energy, the special functional form under consideration gives

$$W_L + W_c = \tfrac{1}{2}\{\mathbf{X}^T\, \mathbf{X} e^{2st} + (\mathbf{X}^T\mathbf{X})^*\, e^{2s^*T} + 2|\mathbf{X}|^T|\mathbf{X}|e^{2\mathrm{Re}[s]t}\} \tag{6-42}$$

where $|\mathbf{X}|$ means a vector whose components are the magnitudes of the complex components of **X**.

Also

$$\begin{aligned} P = {} & (\mathbf{V}^T\, \underset{\sim}{H}\, \mathbf{V})\, e^{2st} + (\mathbf{V}^T\, \underset{\sim}{H}\, V)^*\, e^{2s^*t} \\ & + 2\, \mathrm{Re}\, [\mathbf{V}^{T*}\, H\, \mathbf{V}]\, e^{2\mathrm{Re}[s]t} \end{aligned} \tag{6-43}$$

From Equation (6-36)

$$P_d = e^{2\mathrm{Re}[s]\,t} \begin{Bmatrix} 2[\mathrm{Re}[\mathbf{V}^{T*}\underset{\sim}{H}\ \mathbf{V}] - \mathrm{Re}[s]\ |\mathbf{X}|^T|\mathbf{X}|] \\ + [\mathbf{V}^T\ \underset{\sim}{H}\ \mathbf{V} + s\ \mathbf{X}^T\ \mathbf{X}]\ e^{2j\,\mathrm{Im}[s]\,t} \\ + [\mathbf{V}^T\ \underset{\sim}{H}\ \mathbf{V} + s\ \mathbf{X}^T\ \mathbf{X}]^*\ e^{-2j\,\mathrm{Im}[s]\,t} \end{Bmatrix} \tag{6-44}$$

If P_d is non-negative, the average value of the term in the braces must be non-negative for all time. The first two terms are constants, while the last two together are a sinusoid of frequency $2\ \mathrm{Im}[s]$. Thus the constant terms together must be positive and at least as large as the peak value of the sinusoidal terms.

***PR* transfer function matrices.** The above condition is sufficient to guarantee that $H(s)$ is a positive real matrix defined as follows:

A $p \times p$ matrix $\underset{\sim}{H}$ of complex valued functions of a complex variables is **Positive Real** (PR) if for all complex valued p vectors **V**

$$\mathrm{Re}\ [\mathbf{V}^{T*}\ \underset{\sim}{H}\ \mathbf{V}] \geqslant 0 \qquad \text{for } \mathrm{Re}[s] \geqslant 0$$

and if $\underset{\sim}{H}(\sigma)$, σ real, has all real entries.

Since the systems we are dealing with are characterized by linear differential equations with real coefficients, a real excitation such as $e^{\sigma t}$ gives a real response. Thus the second part of the definition of *PR* is satisfied. When Re[*s*] is positive, P_d positive semidefinite implies

$$\mathrm{Re}[\mathbf{V}^{T*} \underset{\sim}{H} \mathbf{V}] > \mathrm{Re}[s]\ |\mathbf{X}|^T|\mathbf{X}| \geq 0 \tag{6-45}$$

The above derivation has shown that a system with positive semidefinite energy functions W_L, W_C, and P_d, is passive, and its transfer function matrix is *PR*. Thus all *RLC* circuits with positive elements have *PR* matrix characterization. The corresponding statement applies to the mechanical *MKB* circuits. In Chapter 8 we shall investigate some further properties of systems with *PR* matrix descriptions. Before proceeding let us consider some examples.

A mechanical example. As a first example consider the rotational mechanical system of Figure 6-4. The differential gear as drawn is

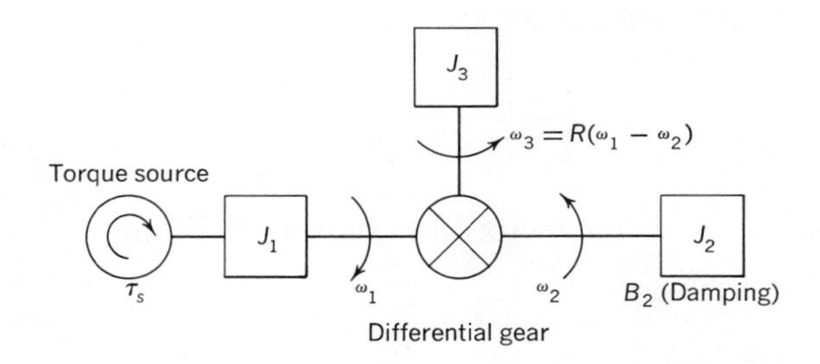

Figure 6-4

frictionless and massless. The masses are taken as parts of the shaft inertia loads, J_1, J_2, and J_3. The friction is assumed to be included in the load damping B_2. Newton's law for each moment of inertia gives

$$\begin{aligned} J_1\dot{\omega}_1 &= -\tau_1 + \tau_s \\ J_2\dot{\omega}_2 &= -\tau_2 - B_2\omega_2 \\ J_3\dot{\omega}_3 &= -\tau_3 \end{aligned} \tag{6-46}$$

where τ_1, τ_2, and τ_3 are the reaction torques of the differential gear on the respective shafts.

Since the differential gear is lossless, power into the gear must equal power out of the gear. Symbolically

$$\omega_1\tau_1 + \omega_2\tau_2 + \omega_3\tau_3 = 0 \tag{6-47}$$

Since the torques must balance in the gear (see Reference 10)

$$\tau_1 = -\tau_2 = -R\ \tau_3 \tag{6-48}$$

By using Equation (6-48) and the defining equation for the differential gear, we can reduce (6-46) to two differential equations with ω_1 and ω_2 as variables. They are

$$\begin{aligned} (J_1 + R^2J_3)\,\dot{\omega}_1 - R^2J_3\,\dot{\omega}_2 &= \tau_s \\ -R^2J_3\,\dot{\omega}_1 + (J_2 + R^2J_3)\,\dot{\omega}_2 &= -B\,\omega_2 \end{aligned} \tag{6-49}$$

These two equations must be put in state variable form with the variables chosen so that the stored kinetic energy is $\frac{1}{2}\mathbf{x}^T\mathbf{x}$.

In terms of the ω's the kinetic energy is

$$\begin{aligned} T &= \tfrac{1}{2}\,[J_1\omega_1^2 + J_2\omega_2^2 + J_3\omega_3^2] \\ &= \tfrac{1}{2}\,[J_1 + R^2J_3)\,\omega_1^2 + (J_2 + R^2J_3)\,\omega_2^2 - 2R^2J_3\,\omega_1\,\omega_2] \\ &= \tfrac{1}{2}\,[\omega_1\ \omega_2]\begin{bmatrix} J_1 + R^2J_3 & -R^2J_3 \\ -R^2J_3 & J_2 + R^2J_3 \end{bmatrix}\begin{bmatrix} \omega_1 \\ \omega_2 \end{bmatrix} \end{aligned} \tag{6-50}$$

The quadradic form (6-50) can be diagonalized just as the form (6-10) was above. The transformation matrix is

$$\underset{\sim}{K} = \begin{bmatrix} 1 & 0 \\ \dfrac{R^2J_3}{J_2 + R^2J_3} & 1 \end{bmatrix} \tag{6-51}$$

The substitution for the ω vector in Equation (6-50) is

$$\omega = \underset{\sim}{K}\,\lambda \tag{6-52}$$

The resulting quadratic form is

$$T = \tfrac{1}{2}\,[\lambda_1 \quad \lambda_2]\begin{bmatrix} J_1 + R^2J_3 - \dfrac{R^4J_3^2}{J_2 + R^2J_3} & 0 \\ 0 & J_2 + R^2J_3 \end{bmatrix}\begin{bmatrix} \lambda_1 \\ \lambda_2 \end{bmatrix} \tag{6-53}$$

If we now define x_1 and x_2 by

$$\begin{aligned} x_1 &= \sqrt{J_1 + R^2J_3 - \frac{R^4J_3^2}{J_2 + R^2J_3}}\ \lambda_1 \\ x_2 &= \sqrt{J_2 + R^2J_3}\ \lambda_2 \end{aligned} \tag{6-54}$$

Then the quadratic form becomes

$$T = \tfrac{1}{2}\,\mathbf{x}^T\mathbf{x} \tag{6-55}$$

The change of variables from ω to $\mathbf{x}$ is the combination of Equations (6-51) and (6-54). It is

$$\omega_1 = \sqrt{\frac{J_2 + R^2J_3}{J_1J_2 + R^2J_3\,(J_1 + J_2)}}\ x_1$$

$$\omega_2 = \frac{R^2 J_3}{J_2 + J_3} \sqrt{\frac{J_2 + R^2 J_3}{J_1 J_2 + R^2 J_3 (J_1 + J_2)}}\, x_1 + \frac{1}{\sqrt{J_2 + R^2 J_3}}\, x_2 \tag{6-56}$$

For the remaining steps of the reduction of (6-49) to normal form in the x variables and the demonstration that the energy formulas apply, there is no educational value to carrying long expressions such as (6-56). To simplify the computations let us choose values for the parameters as follows:

$$J_1 = J_2 = J_3 = R = B_2 = 1 \tag{6-57}$$

With these values Equations (6-49) becomes

$$\begin{aligned} 2\dot{\omega}_1 - \dot{\omega}_2 &= \tau_s \\ -\dot{\omega}_1 + 2\dot{\omega}_2 &= -\omega_2 \end{aligned} \tag{6-58}$$

The transformation (6-56) to the x variables becomes

$$\begin{aligned} \omega_1 &= \sqrt{\frac{2}{3}}\, x_1 \\ \omega_2 &= \frac{1}{\sqrt{6}}\, x_1 + \frac{1}{\sqrt{2}}\, x_2 \end{aligned} \tag{6-59}$$

The reduction steps now proceed as follows. Substituting Equations (6-59) into Equations (6-58) gives

$$\begin{aligned} \left(2\sqrt{\frac{2}{3}} - \frac{1}{\sqrt{6}}\right) \dot{x}_1 - \frac{1}{\sqrt{2}}\, \dot{x}_2 &= \tau_s \\ \left(\frac{2}{\sqrt{6}} - \sqrt{\frac{2}{3}}\right) \dot{x}_1 + \sqrt{2}\, \dot{x}_2 &= -\frac{1}{\sqrt{6}} x_1 - \frac{1}{\sqrt{2}} x_2 \end{aligned} \tag{6-60}$$

Solving Equations (6-60) for $\dot{x}_1$ and $\dot{x}_2$ gives

$$\begin{aligned} \dot{x}_1 &= -\frac{1}{6} x_1 - \frac{1}{2\sqrt{3}} x_2 + \sqrt{\frac{2}{3}}\, \tau_s \\ \dot{x}_2 &= -\frac{1}{2\sqrt{3}} x_1 - \frac{1}{2} x_2 \end{aligned} \tag{6-61}$$

Then input power is

$$P = \omega_1 \tau_s = \sqrt{\frac{2}{3}}\, x_1\, \tau_s \tag{6-62}$$

The derivative of the stored kinetic energy is

$$\dot{T} = \mathbf{x}^T\dot{\mathbf{x}} = -\frac{1}{6}x_1^2 - \frac{1}{2\sqrt{3}}x_1x_2 + \sqrt{\frac{2}{3}}\,\tau_s$$
$$-\frac{1}{2\sqrt{3}}x_1x_2 - \frac{1}{2}x_2 \tag{6-63}$$
$$= -\frac{1}{6}x_1^2 - \frac{1}{\sqrt{3}}x_1x_2 - \frac{1}{2}x_2^2 + \sqrt{\frac{2}{3}}\,\tau_s$$

The dissipative power is

$$P_d = P - \dot{T} = \frac{1}{6}x_1^2 + \frac{1}{\sqrt{3}}x_1x_2 + \frac{1}{\sqrt{2}}x_2^2 \tag{6-64}$$

Since the only dissipative element in the system of Figure 6-4 is the damping on shaft 2, the dissipation should also be

$$P_d = B_2\omega_2^2$$
$$= \left(\frac{1}{\sqrt{6}}x_1 + \frac{1}{\sqrt{2}}x_2\right)^2 \tag{6-65}$$
$$= \frac{1}{6}x_1^2 + \frac{1}{\sqrt{3}}x_1x_2 + \frac{1}{2}x_2^2$$

Thus the results check.

An electrical example. As a second example let us consider the circuit of Figure 6-5. The loop of capacitors gives a complexity equivalent

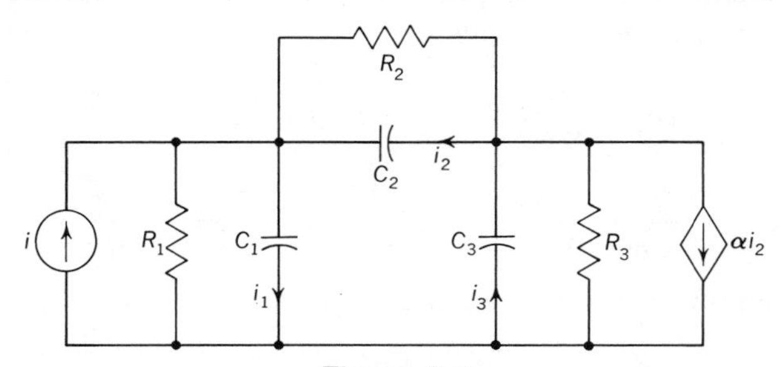

Figure 6-5

to the preceding example. The controlled source with control current and capacitor current adds complexity. With the signs for the variables as indicated by the arrows on the capacitors and the passive sign convention, straightforward application of the methods of Chapter 2 gives

$$C_1\dot{v}_1 = -\frac{1}{R_1}v_1 - \frac{1}{R_3}(v_1 + v_2) + i_3 - \alpha\, C_2\dot{v}_2 + i$$
$$C_2\dot{v}_2 = -\frac{1}{R_2}v_2 - \frac{1}{R_3}(v_1 + v_2) + i_3 - \alpha\, C_2\dot{v}_2 \tag{6-66}$$

Since

$$i_3 = -C_3\ (\dot{v}_1 + \dot{v}_2) \tag{6-67}$$

Equations (6-66) can be regrouped to put derivatives on the left as follows:

$$\begin{aligned} (C_1 + C_3)\ \dot{v}_1 + (C_3 + \alpha\ C_2)\ \dot{v}_2 &= -\left(\frac{1}{R_1} + \frac{1}{R_3}\right) v_1 - \frac{1}{R_3} v_2 + i \\ C_3\dot{v}_1 + (C_2(1+\alpha) + C_3)\ \dot{v}_2 &= -\frac{1}{R_3} v_1 - \left(\frac{1}{R_2} + \frac{1}{R_3}\right) v_2 \end{aligned} \tag{6-68}$$

The stored energy is given by

$$W_c = \tfrac{1}{2}\ \mathbf{v}^T\ \underset{\sim}{C}\ \mathbf{v} \tag{6-69}$$

where

$$\underset{\sim}{C} = \begin{bmatrix} C_1 + C_3 & C_3 \\ C_3 & C_2 + C_3 \end{bmatrix}$$

By following the procedure of the previous example, we can find a matrix K such that

$$\begin{aligned} \mathbf{v} &= \underset{\sim}{K}\ \mathbf{x} \\ W_c &= \tfrac{1}{2}\ \mathbf{x}^T\mathbf{x} \end{aligned} \tag{6-70}$$

This transformation is equivalent to (6-52) and (6-54) of the previous example. If the $\underset{\sim}{C}$ matrix generates a positive definite quadratic form, then $\underset{\sim}{K}$ is nonsingular.

With (6-70) the differential equations (6-68) can be converted to an equation in $\mathbf{x}$ as follows:

$$\underset{\sim}{C}_1\ \underset{\sim}{K}\ \dot{\mathbf{x}} = -\underset{\sim}{G}\ \underset{\sim}{K}\ \mathbf{x} + \begin{bmatrix} i \\ 0 \end{bmatrix} \tag{6-71}$$

where

$$\underset{\sim}{C}_1 = \begin{bmatrix} C_1 + C_3 & C_3 + \alpha C_2 \\ C_3 & C_2\ (1+\alpha) + C_3 \end{bmatrix}$$

$$\underset{\sim}{G} = \begin{bmatrix} \left(\dfrac{1}{R_1} + \dfrac{1}{R_3}\right) & \dfrac{1}{R_3} \\ \dfrac{1}{R_3} & \dfrac{1}{R_2} + \dfrac{1}{R_3} \end{bmatrix}$$

Since $\underset{\sim}{K}$ is nonsingular, (6-71) can be solved for $\mathbf{x}$ so long as $\underset{\sim}{C}_1$ is nonsingular. The determinant of $\underset{\sim}{C}_1$ is

$$|\underset{\sim}{C}_1| = C_1C_2\ (1+\alpha) + C_3\ (C_1 + C_2)$$

If the C_i are all positive, only a particular negative value of α makes the matrix singular. In almost all cases (6-71) can be rewritten as

$$\dot{\mathbf{x}} = \underset{\sim}{K}^{-1} \underset{\sim}{C}_1^{-1} \underset{\sim}{G} \underset{\sim}{K} \mathbf{x} + \underset{\sim}{K}^{-1} \underset{\sim}{C}_1^{-1} \begin{bmatrix} i \\ 0 \end{bmatrix} \tag{6-72}$$

The input power to the system is

$$P = v_1 i = k_{11} x_1 \, i + k_{12} x_2 \, i \tag{6-73}$$

where k_{11} and k_{12} are respective elements of $\underset{\sim}{K}$. The dissipative power is

$$P_d = P + \mathbf{x}^T \underset{\sim}{K}^{-1} \underset{\sim}{C}_1^{-1} \underset{\sim}{G} \underset{\sim}{K} \mathbf{x} - \mathbf{x}^T \underset{\sim}{K}^{-1} \underset{\sim}{C}_1^{-1} \begin{bmatrix} i \\ 0 \end{bmatrix} \tag{6-74}$$

The dissipative power is also the power delivered to the three resistors plus the power to the controlled source. That is,

$$\begin{aligned} P_d = v_1^2 \left(\frac{1}{R_1}\right) + v_2^2 \left(\frac{1}{R_2}\right) + (v_1 + v_2)^2 \frac{1}{R_3} \\ + (v_1 + v_2) \, (\alpha \, C_2 \, \dot{v}_2) \end{aligned} \tag{6-75}$$

In order to compare (6-75) with (6-74) v_1 and v_2 must be written in terms of x_1 and x_2 by (6-70). This equation also permits us to write

$$\dot{v}_2 = k_{21} \, \dot{x}_1 + k_{22} \, \dot{x}_2 \tag{6-76}$$

Next, Equation (6-76) must be written in terms of x_1, x_2, and i by using (6-72). This result can be used in (6-75) to get P_d in terms of x_1, x_2, and i. The details are left as an exercise for the student.

PROBLEMS

6-1 For the circuit shown in Figure P6-1 with all resistors $1\,\Omega$, find the range of values of g_m such that the 2-port is passive. Select a value of g_m outside of the passive range and find a pair of inputs such that the circuit delivers net power at its ports.

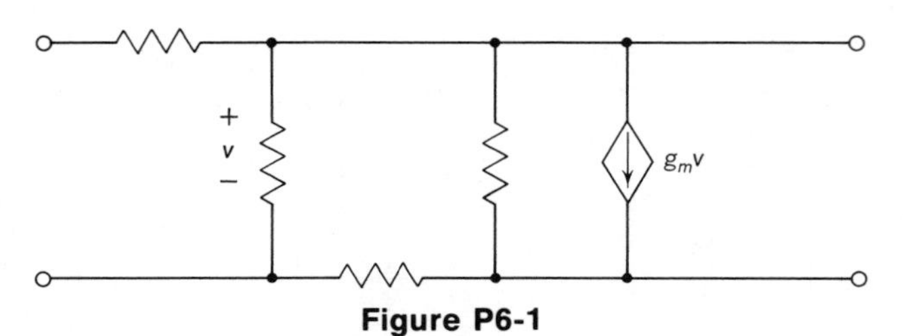

Figure P6-1

6-2 a. Show that two passive, resistive 2-ports cascaded with an ideal transformer as shown in Figure P6-2(a) is a passive 2-port.

b. Show that two passive, resistive 2-ports cascaded with a gyrator as shown in Figure P6-2(b) is a passive 2-port.

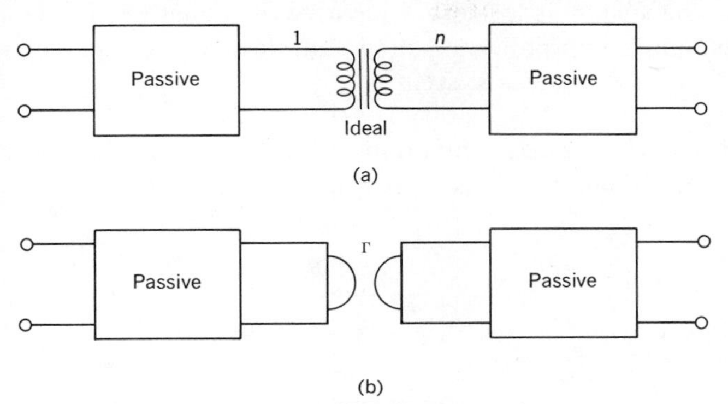

Figure P6-2

6-3 Show that the circuit of Figure P6-3 is lossless. Then set up the normal form equations with variables chosen so that $\frac{1}{2}\mathbf{x}^T\mathbf{x}$ is the total stored energy and $\mathbf{x}$ is two-dimensional.

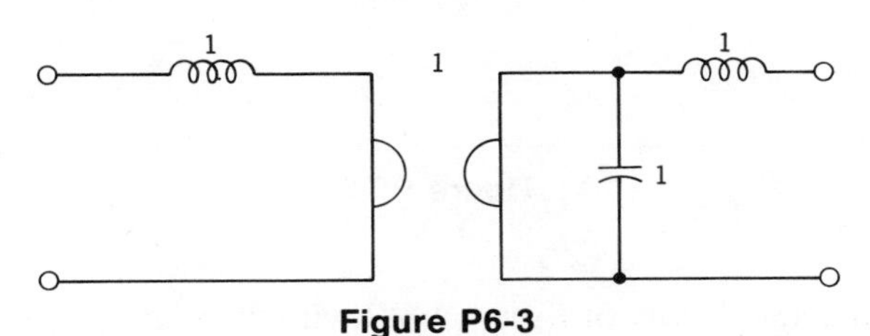

Figure P6-3

6-4 Set up the equations of motion for the cylindrical pendulum whose end view is shown in Figure P6-4. The small cylinder of mass M rolls without slipping on the fixed larger cylinder. The force of gravity pulls the small cylinder toward the bottom of the larger one. Compute the kinetic and potential energies and show that the system is lossless.

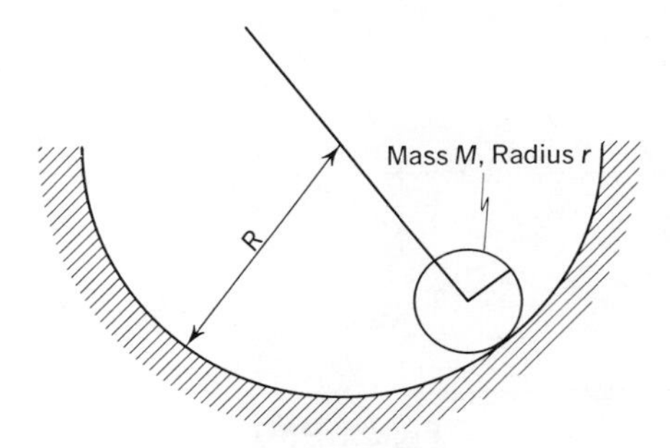

Figure P6-4

6-5 Consider the mechanical system of Problem 1-8 with no frictional force. Set up the equation of motion and compute the energy functions. Show that the system is lossless. Linearize the equations for small notion about the equilibrium point and show that these linear equations are also the equations of a lossless system.

6-6 Construct the energy functions W_L, W_C, and P_d for the circuit of Figure P6-6. Define variables in any appropriate fashion.

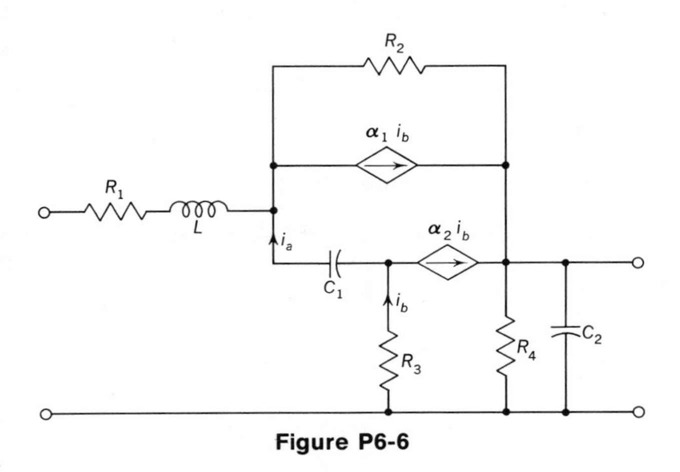

Figure P6-6

6-7 Consider the 2-port of Figure P6-7 with current source input at the left port and voltage source input on the right.

a. Set up equations in the form (6-1).

b. Select a set of values for the parameters so that P_d, W_L, and W_c are positive (semi) definite.

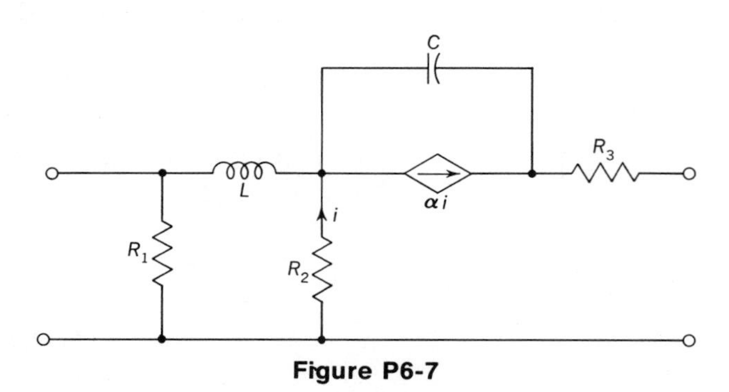

Figure P6-7

6-8 Set up normal form equations for the circuit of Figure P6-8 such that $\frac{1}{2}\mathbf{x}^T\mathbf{x}$ is the total stored energy with $\mathbf{x}$ two-dimensional.

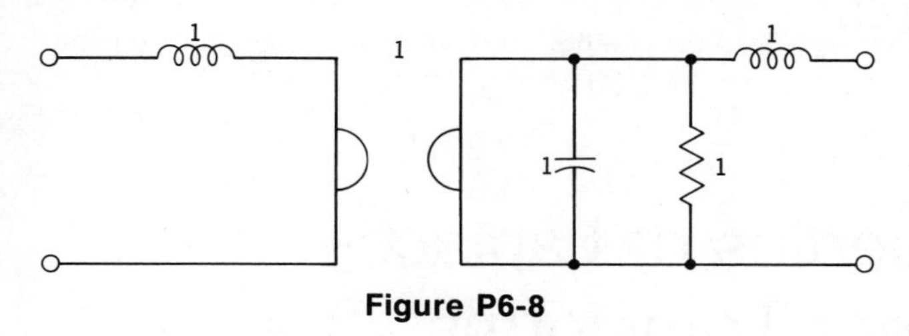

Figure P6-8

CHAPTER

7

Properties of Laplace-Fourier Transforms

In the previous chapter we saw that systems with positive semidefinite energy and dissipation functions were characterized by *PR* matrixes. In the next chapter we shall see that any system characterized by a *PR* transfer function matrix has certain important engineering properties. We shall also see that the class of *PR* systems is larger than the class of systems with *PSD* energy and dissipation functions.

In order to demonstrate the properties of *PR* systems we shall need to use some important properties of Laplace-Fourier transforms. The student is assumed to have worked with transforms at the undergraduate level as part of an intermediate circuits or systems course or in a differential equations course. The level assumed is slightly less than that of Reference 6 or Reference 11. In this brief chapter we shall discuss some of the properties of Laplace-Fourier transforms that are important for Chapters 8 and 10 but are not usually emphasized in undergraduate studies.

7-1 Basic Properties of Two-Sided Transforms

In order to account for all types of signals and signal processing in the systems under consideration we use the two-sided transform defined for those values of s wherein the integral exists by

$$F(s) = \int_{-\infty}^{\infty} f(t)\, e^{-st}\, dt \tag{7-1}$$

The inverse transform is

$$f(t) = \frac{1}{2\pi j} \int_{\sigma - j\infty}^{\sigma + j\infty} F(s)\, e^{st}\, ds \tag{7-2}$$

for the same values of s as allowed above.

When we use the word existence for the integrals we must include the possibility of δ functions and their derivatives in both t and s domains. If

$$f(t) = \delta(t); \qquad F(s) = 1 \tag{7-3}$$

if

$$F(s) = \delta(s); \qquad f(t) = \frac{1}{2\pi} \tag{7-4}$$

The problems involved with relating the appropriate regions of the s-plane with the transform pair (7-1) and (7-2) is best illustrated by an example.

Suppose

$$f(t) = e^{-a|t|} \tag{7-5}$$

a real and positive.

Then by (7-1)

$$F(s) = \int_{-\infty}^{0} e^{at}\, e^{-st}\, dt + \int_{0}^{\infty} e^{-at}\, e^{-st}\, dt \tag{7-6}$$

If $-a < \text{Re}\,[s] < a$, both integrals in (7-6) are well-defined and

$$F(s) = \frac{1}{s+a} - \frac{1}{s-a} \tag{7-7}$$

Now consider the function

$$g(t) = \begin{cases} 0 & \text{for } t < 0 \\ e^{-at} - e^{at} & \text{for } t \geqslant 0 \end{cases} \tag{7-8}$$

Then

$$G(s) = \int_{0}^{\infty} e^{-at}\, e^{-st}\, dt - \int_{0}^{\infty} e^{at}\, e^{-st}\, dt \tag{7-9}$$

For $\text{Re}\,[s] > a$, both integrals converge and

$$G(s) = \frac{1}{s+a} - \frac{1}{s-a} \tag{7-10}$$

Thus, the time functions (7-5) and (7-8) both correspond to the same frequency function (7-7) and (7-10). The difference is that they correspond to it in different nonoverlapping regions of the s-plane. There is a unique relation between frequency and time function provided the region of the s-plane, where the two are a transform pair, is included.

When dealing with the Laplace-Fourier transform—sometimes called the two-sided Laplace transform or the complex Fourier transform—one must be careful to specify the region of the s-plane along with the appropriate frequency function. The above example shows that a right half plane pole may mean a function that grows exponentially or it could mean a function that is nonzero for negative time.

7-2 Even and Odd Functions

If $f(t)$ is even and $F(j\omega)$ exists, then $F(j\omega)$ is real. If $f(t)$ is odd, $F(j\omega)$ is imaginary if it exists. When $f(t) = 0$ for $t < 0$, we can define

$$f_e(t) = \begin{cases} \frac{1}{2} f(t) & \text{for } t > 0 \\ f(t) & \text{for } t = 0 \\ \frac{1}{2} f(-t) & \text{for } t < 0 \end{cases} \tag{7-11}$$

and

$$f_0(t) = \begin{cases} \frac{1}{2} f(t) & \text{for } t > 0 \\ 0 & \text{for } t = 0 \\ -\frac{1}{2} f(-t) & \text{for } t < 0 \end{cases} \tag{7-12}$$

Now f_e is even, f_0 odd, and

$$f(t) = f_e(t) + f_0(t) \tag{7-13}$$

$F(j\omega) = R(\omega) + jX(\omega)$ with

$$\int_{-\infty}^{\infty} f_e(t)\, e^{-j\omega t}\, dt = R(\omega) \tag{7-14}$$

$$\int_{-\infty}^{\infty} f_0(t)\, e^{-j\omega t}\, dt = jX(\omega) \tag{7-15}$$

The inverses are

$$f_e(t) = \frac{1}{2\pi} \int_{-\infty}^{\infty} R(\omega)\, e^{j\omega t}\, d\omega \tag{7-16}$$

$$= \frac{2}{2\pi} \int_0^\infty R(\omega) \cos \omega t \, d\omega$$

since R is even and $e^{j\omega t} = \cos \omega t + j \sin \omega t$. Finally,

$$f(t) = \frac{2}{\pi} \int_0^\infty R(\omega) \cos \omega t \, d\omega \qquad \text{for } t \geqslant 0 \tag{7-17}$$

Similarly,

$$f(t) = -\frac{2}{\pi} \int_0^\infty X(\omega) \sin \omega t \, d\omega \qquad \text{for } t > 0 \tag{7-18}$$

Thus, when a time function is zero for $t < 0$, either the real or imaginary part of the transform is sufficient to determine the function. The integral relationships between $R(\omega)$ and $f(t)$ and between $X(\omega)$ and $f(t)$ are called the *Fourier cosine* and *sine transform*, respectively.

Since $f(t)$ can be determined from $R(\omega)$, and $X(\omega)$ can be determined from $f(t)$, a formula for computing $X(\omega)$ from $R(\omega)$ can be derived when $f(t) = 0$ for $t < 0$. Many interesting time functions, including some needed to derive the desired formula, are such that the direct transform integral (7-14) does not exist in the usual sense for $s = j\omega$. Thus, before deriving the relationship between $R(\omega)$ and $X(\omega)$ we must show how certain integrals can be considered by using delta functions along with functions in the usual sense. The derivation given herein is not rigorous; rather, it is a sequence of heuristic steps that gives the right answer.

The unit step function, $U(t)$, has a transform $1/s$ for Re $[s] > 0$. Let us consider $1/j\omega$ as the imaginary part of the transform of a time function that is zero for $t < 0$. If the time function is still the unit step, then its odd part is $\frac{1}{2}[U(t) - U(-t)]$, the unit odd function. Its even part is $\frac{1}{2}[U(t) + U(-t)]$. This even part is the constant $\frac{1}{2}$ except at the point $t = 0$ where it takes on the value of 1. Such an isolated point has no effect on the transform integral. From (7-4), the transform of $\frac{1}{2}$ is $\pi\delta(s)$. Thus, for $s = j\omega$, the transform of the unit step function is $\pi\delta(j\omega) - j(1/\omega)$. Thus, $R(\omega) = \pi\delta(\omega)$; and $X(\omega) = -1/\omega$.

The above reasoning can be checked by computing the inverse transform from (7-2).

$$\begin{aligned} f(t) &= \frac{1}{2\pi j} \int_{-j\infty}^{j\infty} \left[\pi\delta(j\omega) + \frac{1}{j\omega} \right] e^{j\omega t} \, dj\omega \\ &= \frac{1}{2\pi} \int_{-\infty}^{\infty} \left[\pi\delta(\omega) - j\frac{1}{\omega} \right] [\cos \omega t + j \sin \omega t] \, d\omega \end{aligned} \tag{7-19}$$

Now

$$\int_{-\infty}^{\infty} \delta(\omega) \cos \omega t \, d\omega = 1 \tag{7-20}$$

$$\int_{-\infty}^{\infty} \delta(\omega) \sin \omega t \, d\omega = 0 \tag{7-21}$$

$$\int_{-\infty}^{\infty} \frac{\cos \omega t}{\omega} \, d\omega = 0 \tag{7-22}$$

$$\int_{-\infty}^{\infty} \frac{\sin \omega t}{\omega} \, d\omega = \begin{cases} -\pi & \text{if } t < 0 \\ 0 & \text{if } t = 0 \\ \pi & \text{if } t > 0 \end{cases} \tag{7-23}$$

Thus,

$$f(t) = \left\{ \begin{matrix} 0 & \text{if } t < 0 \\ \frac{1}{2} & \text{if } t = 0 \\ 1 & \text{if } t > 0 \end{matrix} \right\} = U(t) \tag{7-24}$$

By a similar argument to the above, or by use of the shifting theorem for transforms, one can easily show that for $s = j\omega$, the transform of

$$f(t) = \begin{cases} e^{j\omega_0 t} & \text{for } t \geq 0 \\ 0 & \text{for } t < 0 \end{cases} \tag{7-25}$$

is

$$F(j\omega) = \pi\delta(j(\omega - \omega_0)) + \frac{1}{j(\omega - \omega_0)} \tag{7-26}$$

Transforms of convolutions. On the j axis, where the Laplace-Fourier transform, when it exists, is in a strict sense a Fourier transform, convolution in the time domain corresponds to multiplication in the frequency domain, and vice versa. Consider the transform of the convolution of two time functions.

Let

$$f(t) = \int_{-\infty}^{\infty} f_1 \, (t - \tau) \, f_2(\tau) \, d\tau \tag{7-27}$$

Then

$$F(j\omega) = \int_{-\infty}^{\infty} \int_{-\infty}^{\infty} f_1(t - \tau) \, f_2(\tau) \, e^{-j\omega t} \, d\tau \, dt \tag{7-28}$$

The change of variables $\lambda = t - \tau$ in the t integration gives

$$\begin{aligned} F(j\omega) &= \int_{-\infty}^{\infty} \int_{-\infty}^{\infty} f_1(\lambda) \, f_2(\tau) \, e^{-j\omega(\lambda+\tau)} \, d\lambda \, d\tau \\ &= \int_{-\infty}^{\infty} f_1(\lambda) \, e^{-j\omega\lambda} \, d\lambda \int_{-\infty}^{\infty} f_2(\tau) \, e^{-j\omega\tau} \, d\tau \\ &= F_1(j\omega) \, F_2(j\omega) \end{aligned} \tag{7-29}$$

For convolution in the frequency domain the starting point is

$$F(j\omega) = \int_{-\infty}^{\infty} F_1(j(\omega - \lambda))\, F_2(j\lambda)\, d\lambda \tag{7-30}$$

Then

$$f(t) = \frac{1}{2\pi} \int_{-\infty}^{\infty} \int_{-\infty}^{\infty} F_1(j(\omega - \lambda))\, F_2(j\lambda)\, e^{j\omega t}\, d\lambda\, d\omega \tag{7-31}$$

The change of variables $\sigma = \omega - \lambda$ in the ω integration gives

$$f(t) = \left[\frac{1}{2\pi} \int_{-\infty}^{\infty} F_1(j\sigma)\, e^{j\sigma t}\, d\sigma \right] \left[\int_{-\infty}^{\infty} F_2(j\lambda)\, e^{j\lambda t}\, d\lambda \right] \tag{7-32}$$

Thus

$$f(t) = 2\pi\, f_1(t)\, f_2(t) \tag{7-33}$$

In summary the transform pair relationship is

$$\int_{-\infty}^{\infty} f_1(t - \tau)\, f_2(\tau)\, d\tau \longleftrightarrow F_1(j\omega)\, F_2(j\omega) \tag{7-34}$$

$$f_1(t)\, f_2(t) \longleftrightarrow \frac{1}{2\pi} \int_{-\infty}^{\infty} F_1(j(\omega - \lambda))\, F_2(j\lambda)\, d\lambda \tag{7-35}$$

The Hilbert transform. With the above special transform pairs we can return to the relation between real and imaginary parts of the transforms of time functions that are zero for $t < 0$. From Equations (7-15), (7-12), (7-11), and (7-16)

$$\begin{aligned} jX(\omega) &= \int_{-\infty}^{\infty} f_0(t)\, e^{-j\omega t}\, dt \\ &= \int_{-\infty}^{0} \frac{-1}{2} f(-t)\, e^{-j\omega t}\, dt + \int_{0}^{\infty} f_e(t)\, e^{-j\omega t}\, dt \\ &= \int_{-\infty}^{0} -f_e(t)\, e^{-j\omega t}\, dt + \int_{0}^{\infty} f_e(t)\, e^{-j\omega t}\, dt \\ &= -\int_{-\infty}^{0} \left[\frac{1}{2\pi} \int_{-\infty}^{\infty} R(\lambda)\, e^{j\lambda t}\, d\lambda \right] e^{-j\omega t}\, dt \\ &\quad + \int_{0}^{\infty} \left[\frac{1}{2\pi} \int_{-\infty}^{\infty} R(\lambda)\, e^{j\lambda t}\, d\lambda \right] e^{-j\omega t}\, dt \end{aligned} \tag{7-36}$$

Assuming that the order of integration can be changed, a step that should really be justified because of the infinite limits, we have

$$\int_0^\infty \left[\int_{-\infty}^\infty R(\lambda)\, e^{j\lambda t}\, d\lambda \right] e^{-j\omega t}\, dt = \int_{-\infty}^\infty R(\lambda) \left[\int_0^\infty e^{j\lambda t}\, e^{-j\omega t}\, dt \right] d\lambda \qquad \text{(7-37)}$$

But the integral in the bracket is the Laplace-Fourier transform of $U(t)e^{j\lambda t}$. We saw above that this transform is $\pi\delta(j(\omega-\lambda)) + 1/j(\omega-\lambda)$. A similar argument shows that

$$\int_{-\infty}^0 e^{j\lambda t}\, e^{-j\omega t}\, dt = \pi\delta(j(\lambda-\omega)) + \frac{1}{j(\lambda-\omega)} \qquad \text{(7-38)}$$

Thus,

$$jX(\omega) = \frac{1}{2\pi}\int_{-\infty}^\infty R(\lambda) \left\{ \pi\delta(j(\omega-\lambda)) + \frac{1}{j(\omega-\lambda)} - \pi\delta(j(\lambda-\omega)) + \frac{1}{j(\omega-\lambda)} \right\} d\lambda \qquad \text{(7-39)}$$

Finally,

$$X(\omega) = -\frac{1}{\pi}\int_{-\infty}^\infty \frac{R(\lambda)}{\omega-\lambda}\, d\lambda \qquad \text{(7-40)}$$

Formula (7-40) is known as the *Hilbert transform*. By a corresponding derivation starting from (7-14), the formula for $R(\omega)$, the inverse transform relation, is found to be

$$R(\omega) = \frac{1}{\pi}\int_{-\infty}^\infty \frac{X(\lambda)}{\omega-\lambda}\, d\lambda \qquad \text{(7-41)}$$

Sometimes the Hilbert transform integrals are in slightly different forms. Since $R(\omega)$ is even,

$$\begin{aligned} X(\omega) &= -\frac{1}{\pi}\int_0^\infty \left(\frac{R(\lambda)}{\omega-\lambda} + \frac{R(\lambda)}{\omega+\lambda} \right) d\lambda \\ &= -\frac{1}{\pi}\int_0^\infty \frac{2\omega R(\lambda)}{\omega^2-\lambda^2}\, d\lambda \\ &= -\frac{2\omega}{\pi}\int_0^\infty \frac{R(\lambda)}{\omega^2-\lambda^2}\, d\lambda \end{aligned} \qquad \text{(7-42)}$$

Similarly, since $X(\omega)$ is odd,

$$R(\omega) = \frac{2}{\pi}\int_0^\infty \frac{\lambda X(\lambda)}{\omega^2-\lambda^2}\, d\lambda \qquad \text{(7-43)}$$

The above discussion demonstrated that if $f(t) = 0$ for $t < 0$, the corresponding $R(\omega)$ and $X(\omega)$ are related by the Hilbert transform, provided that the Laplace-Fourier transform of $f(t)$ is defined in the sense considered herein. The implication is also true the other way; that is, given a frequency function $F(s)$ with $R(\omega)$ and $X(\omega)$ defined and related by the Hilbert transform, the corresponding time function is zero for $t < 0$. The demonstration of this statement proceeds as follows:

Given $R(\omega)$ and $X(\omega)$, $R(\omega)$ even, $X(\omega)$ odd, and R and X related by the Hilbert transform pair (7-40) and (7-41). The corresponding time function is

$$f(t) = \frac{1}{2\pi} \int_{-\infty}^{\infty} [R(\omega) + jX(\omega)]\, e^{j\omega t}\, d\omega \tag{7-44}$$

Any time function can be separated into even and odd parts. By the nature of R and X, it turns out that the even part is

$$f_e(t) = \tfrac{1}{2}[f(t) + f(-t)] = \frac{1}{2\pi} \int_{-\infty}^{\infty} R(\omega)\, e^{j\omega t}\, d\omega \tag{7-45}$$

The odd part of $f(t)$ is

$$f_0(t) = \tfrac{1}{2}[f(t) - f(-t)] = \frac{j}{2\pi} \int_{-\infty}^{\infty} X(\omega)\, e^{j\omega t}\, d\omega \tag{7-46}$$

Substituting (7-40) into (7-46) gives

$$f_0(t) = \frac{j}{2\pi} \int_{-\infty}^{\infty} \int_{-\infty}^{\infty} \frac{-R(\lambda)}{\pi(\omega - \lambda)}\, e^{j\omega t}\, d\lambda\, d\omega \tag{7-47}$$

The last expression is the inverse transform of the convolution of $R(\omega)$ and $1/j\pi\omega$. We can rewrite it in the form

$$f_0(t) = \frac{1}{2\pi} \int_{-\infty}^{\infty} \left[\frac{1}{2\pi} \int_{-\infty}^{\infty} R(\lambda)\, \frac{2}{j(\omega - \lambda)}\, d\lambda \right] e^{j\omega t}\, d\omega \tag{7-48}$$

In the discussion of special transform pairs above we saw that the inverse Laplace-Fourier transform of $1/j\omega$ is the odd function

$$0(t) = \begin{cases} \frac{1}{2} & \text{for } t > 0 \\ 0 & \text{for } t = 0 \\ -\frac{1}{2} & \text{for } t < 0 \end{cases} \tag{7-49}$$

The inverse transform of $R(\omega)$ is just $f_e(t)$. Thus, by the convolution formula (7-35) we have

$$f_0(t) = 2 f_e(t)\, 0(t) \tag{7-50}$$

But with this relation between even and odd parts, $f(t) = 0$ for $t < 0$.

The above facts about the Hilbert transform can be proven for frequency functions that are analytic in the right half s-plane using the theory of functions of a complex variable (see Reference 12). In addition to the above, it can be shown that any function $F(s)$ that is analytic in the right half s-plane and has the property $F(s^*) = F^*(s)$ has real and imaginary parts for $s = j\omega$ that are related by the Hilbert transform. Thus, any rational function with no right half plane poles is the transform of a time function that is zero for $t < 0$ if the inverse transform exists for $s = j\omega$ and is real.

PROBLEMS

7-1 Show that the formula for the strict sense Laplace transform of a derivative of a function is consistent with the generalized Laplace-Fourier transform that includes impulses, as in Section 7-1. The formula is

$$\int_0^\infty \dot{f}(t)\, e^{-st}\, dt = s \int_0^\infty f(t)\, e^{-st}\, dt - f(0)$$

To correspond with Section 7-1 this formula should be compared with the transform of $(d/dt)\ [f(t)\ U(t)]$ where $U(t)$ is the unit step function.

7-2 Show that the transform of Equations (4-27), (4-29), and (4-30) are consistent. That is, show the transform of the left side is the same as that of the right.

7-3 Find the transforms of (4-31) and (4-32) by integration. Then show that the two results are related by the derivative formula.

7-4 Derive formula (7-41) starting from the fact that $f(t) = 0$ for $t < 0$.

CHAPTER

8

Properties of Lumped Systems Characterized by *PR* Matrices

In Chapter 6 we saw that a lumped system with positive semidefinite energy and dissipation functions is passive and has a *PR* hybrid matrix characterization. In this chapter we consider all systems characterized by *PR* matrixes, called *PR systems*. As shown by example, this class includes some systems whose dissipation functions are not *PSD*. The entire class of *PR* systems is passive, causal, and stable in certain senses as defined below. Furthermore, the real and imaginary parts of all transfer functions are related by the Hilbert transform.

8-1 The Poles and Zeros of *PR* Matrix Elements

The first step in ascertaining the properties of *PR* systems is the location of the poles and zeros of the elements of the *PR* matrix that characterizes the system. Since lumped systems have rational transfer functions, the poles and zeros plus the constant multipliers completely determine the various matrix elements. In this section, we show that the *PR*ness of $\underset{\sim}{H}(s)$ implies:

1. No element of $\underset{\sim}{H}(s)$ has right half plane poles.
2. The diagonal elements of $H(s)$ have no right half plane zeros.
3. The residues of poles on the j-axis (real frequencies) are restricted.

The definition of *PR* is a condition of the quadratic form

$$P_0(s) = \mathbf{V}^{T*} \underset{\sim}{H}(s)\ \mathbf{V} \tag{8-1}$$

where $\mathbf{V}$ is a vector of arbitrary complex numbers and in our case $\underset{\sim}{H}(s)$ is a square matrix of rational functions. If any element of $\underset{\sim}{H}(s)$ has a pole at s_0, then there is a vector $\mathbf{V}$ such that $P_0(s)$ also has a pole at s_0.

To show that there are no right half plane poles in $\underset{\sim}{H}(s)$ we first note the behavior of a rational function in the neighborhood of a pole. If $P_0(s)$ has a simple pole at s_0, then

$$P_0(s) = \frac{K_1}{s - s_0} + Q(s) \tag{8-2}$$

where $Q(s)$ is well-defined and bounded near s_0. Very near s_0, $Q(s)$ is small compared to $K_1/(s - s_0)$. Therefore,

$$P_0(s_0 + \epsilon) \sim \frac{K_1}{\epsilon}$$

with ϵ a small complex number.

If we let $\epsilon = |\epsilon|\ e^{j\theta}$ and take θ from 0 to 2π, we see that $\mathrm{Re}[K_1/\epsilon]$ changes sign twice. But by the definition Re $[P_0(s)]$ cannot change signs in the right half plane. Thus, the first property of *PR* matrix elements given above is proven for simple poles. For a pole of order m there are m sign changes on a small circle about the pole. Thus, the property is proven for higher-order poles as well.

A sign change argument similar to that given for poles can also be made for zeros by considering the power-series expansion about the zero. If a diagonal element, say H_{ii}, has a zero, then the quadratic form has a zero when $V_j = 0$, $j \neq i$, and $V_i \neq 0$. This proves the second statement. Off-diagonal elements can have right half plane zeros since an off-diagonal element appears in the quadratic form only when two diagonal elements also appear.

When there are j-axis poles in $P_0(s)$ the first thing we can say is that the poles must be simple. If there were a pole of order m, then the dominant term would be $K_m/(s - s_0)^m$. This term has $2m$ sign changes spaced $180/m$ degrees apart on a small circle around the pole. Thus, if $m > 1$, there must be at least one negative region in the right half plane. For a simple pole on the j-axis at ω_0, the dominant term is $K/(s - j\,\omega_0)$. If K is real and positive, the negative region around the pole lies entirely in the left half plane. We conclude that the main diagonal elements in $\underset{\sim}{H}(s)$ can have simple poles with positive real residues. For off-diagonal elements, j-axis poles are allowed if their residues satisfy certain constraints similar to the ones worked out below for a 2-port circuit.

***PR* 1-ports.** To be more specific about the poles, zeros, and residues of *PR* matrix elements, let us examine some special cases. For a 1-port the quadratic form is just

$$P_0(s) = \mathbf{V}^{T*} \underset{\sim}{H} \mathbf{V} = Z(s)\ |I|^2 \qquad \text{or} \qquad Y(s)\ |V|^2 \tag{8-3}$$

Dividing by the real positive $|I|^2$ or $|V|^2$ does not disturb the *PR* property. Thus, a *PR* 1-port is characterized by a *PR* driving point impedance or admittance. The poles and zeros of $P_0(s)$ can occur only at poles and zeros of $Z(s)$ or $Y(s)$. Since

$$Z(s) = \frac{1}{Y(s)} \tag{8-4}$$

the poles of one are the zeros of the other, and vice versa.

For the 1-port, the transfer function matrix is 1×1—a function of s. Such a 1×1 *PR* matrix is called a *PR* function. *PR* functions are particularly important because the diagonal elements of any *PR* matrix must be *PR* functions. By Equations (8-3) and (8-4) when a function is *PR*, its reciprocal is also *PR*. Since $Z(s)$ or $Y(s)$ is the ratio of polynomials, the roots of both the numerator and denominator polynomials must lie in the closed left half plane. A polynomial of this type is known as a *Hurwitz polynomial.*[1] The fact that the numerator and denominator of an immitance are both Hurwitz is a necessary condition for *PR*ness. It is not sufficient. We should note that a *PR* immitance cannot have a negative resistance at any real (j-axis) frequency. The example below demonstrates an immitance that has a Hurwitz numerator and denominator, no j-axis poles or zeros, including ∞, and still has a negative resistance over some band of frequencies.

To be more specific let us consider the general rational function with a second-degree numerator and denominator

$$\begin{aligned} Z(s) &= \frac{A_2 s^2 + A_1 s + A_0}{B_2 s^2 + B_1 s + B_0} = \frac{A_2}{B_2} \frac{(s - a_1)(s - a_2)}{(s - b_1)(s - b_2)} \\ &= R(\sigma, \omega) + jX(\sigma, \omega) \end{aligned} \tag{8-5}$$

where

$$s = \sigma + j\omega \qquad \sigma \text{ and } \omega \text{ real}$$

$$R(\sigma, \omega) = \frac{(A_2(\sigma^2 - \omega^2) + A_1\sigma + A_0)(B_2(\sigma^2 - \omega^2) + B_1\sigma + B_0) + \omega^2(A_1 + 2A_2\sigma)(B_1 + 2B_2\sigma)}{(B_2(\sigma^2 - \omega^2) + B_1\sigma + B_0)^2 + \omega^2(B_1 + 2B_2\sigma)^2}$$

$$X(\sigma, \omega) = \frac{\omega[(A_1 + 2A_2\sigma)(B_2(\sigma^2 - \omega^2) + B_1\sigma + B_0) - (B_1 + 2B_2\sigma)(A_2(\sigma^2 - \omega^2) + A_1\sigma + A_0)]}{(B_2(\sigma^2 - \omega^2) + B_1\sigma + B_0)^2 + \omega^2(B_1 + 2B_2\sigma)^2}$$

[1] The term *Hurwitz polynomial* is sometimes restricted to polynomials that have only left half plane roots, excluding the j-axis. The most common engineering usage is to allow j-axis roots in a Hurwitz polynomial and to use the term *strictly Hurwitz* when j-axis roots are to be excluded.

The definition of *PR* immediately places conditions on the first and the third of the forms (8-5) for $Z(s)$. Since $Z(s)$ is real when s is real, the A's and B's in the first form must be real. When this condition is met, $R(\sigma, \omega)$ and $X(\sigma, \omega)$ are respectively the real and imaginary parts of $Z(s)$. The *PR* definition requires $R(\sigma, \omega) \geqslant 0$ for $\sigma \geqslant 0$. Since $Z(\sigma)$ must be non-negative for σ large and positive, A_2 and B_2 must have the same sign. A similar argument at $\sigma = 0$ shows that A_0 and B_0 must have the same sign. Of course, one of the terms A_2, B_2, A_0, or B_0 could be zero. Then the other term with the same subscript must be positive. For definiteness let us assume A_2 and B_2 are both positive. There is no generality lost by this assumption.

When neither A_2 or B_2 is zero, then the factored form, the second form in (8-5), is appropriate. From our previous discussion about poles and zeros, we see that the real parts of the a_i and b_i must be negative. From the quadratic formula it is clear that A_1 must be positive when A_2 is positive and B_1 must be positive when B_2 is. That is,

$$\begin{aligned} a_1 &= -\frac{A_1}{2A_2} + \sqrt{\left(\frac{A_1}{2A_2}\right)^2 - A_2 A_0} \\ a_2 &= -\frac{A_1}{2A_2} - \sqrt{\left(\frac{A_1}{2A_2}\right)^2 - A_2 A_0} \end{aligned} \tag{8-6}$$

If A_1 is negative, then $\text{Re}[a_1] > 0$. Furthermore, if $A_1 > 0$ and $A_0 < 0$, a_1 is real and positive. Thus, A_0 must be positive also.

The fact that the A_i and B_i must all be positive (or all negative) generalizes to rational *PR* functions with any degree numerator and denominator polynomials. The poles and zeros must occur either on the real axis or they must occur in conjugate pairs. A root on the negative real axis gives a factor $(s + A)$ where A is real and positive. A pair of conjugate roots that are in the closed left half plane lead to factors of the form $(s^2 + As + B)$ with A and B both positive. When the factors are multiplied out and regrouped in standard form as sums of powers of s, all coefficients are positive.

The above conditions on the coefficients of the polynomials are necessary for a *PR* function, but they are not sufficient. As an example let us consider

$$\begin{aligned} Z(s) &= \frac{s^2 + 30s + 200}{s^2 + 3s + 2} = \frac{(s+10)(s+20)}{(s+1)(s+2)} \\ &= \frac{(\sigma^2 - \omega^2 + 300 + 200)(\sigma^2 - \omega^2 + 3\sigma + 2) + \omega^2(4\sigma^2 + 66\sigma + 90)}{(\sigma^2 - \omega^2 + 3\sigma + 2)^2 + \omega^2(2\sigma + 3)^2} \\ &\qquad + jX(\sigma, \omega) \end{aligned} \tag{8-7}$$

There is no need to write out $X(\sigma, \omega)$ since *PR*ness places no restriction thereon. For $\sigma = 0$,

$$R(0,\omega)=\frac{(200-\omega^2)(2-\omega^2)+90\omega^2}{(2-\omega^2)^2+9\omega^2}=\frac{\omega^4-112\omega^2+400}{(2-\omega^2)^2+9\omega^2} \quad \textbf{(8-8)}$$

Since the denominator in (8-8) is positive, the possibility of $R(0, \omega)$ being negative depends on the numerator. Taking the derivative with respect to ω^2 and setting the result to zero gives

$$2\,\omega^2 - 112 = 0 \quad \textbf{(8-9)}$$

There is a maximum or minimum at

$$\omega^2 = 56 \quad \textbf{(8-10)}$$

Now

$$R(0, \sqrt{56}) = \frac{-\,2736}{(54)^2 + 9 \times 56} \quad \textbf{(8-11)}$$

Clearly (8-7) is not *PR*.

For a 1-port impedance or admittance function restriction of poles and zeros to the left half plane is necessary but not sufficient for *PR*ness. The restriction on the real part for $\sigma \geqslant 0$ must still be checked. From formula (8-5), it is apparent that the expression for the real part is quite complicated even in the case of second-degree polynomials. Since $R(\sigma, \omega)$ is a function of two variables we should have to compute the gradient and set it to zero to find the minima. Then each must be evaluated to see if there are points with $R(\sigma, \omega) < 0$ in the right half plane.

Fortunately there is a theorem in the theory of analytic functions that narrows the search for the minimum. It is as follows (Reference 12, p. 334):

> If a function of a complex variable is analytic over a region including the boundary, the maximum and minimum values of its real and imaginary parts for that region must occur on the boundary.

In the present case the functions are analytic in the right half plane when there are no right half plane poles. (Rational functions are analytic except at poles.) The boundary is the j-axis. Thus, if a rational function has no right half plane or j-axis poles, then

$$\min R(\sigma, \omega)|_{\sigma \geqslant 0} = \min R(0, \omega) \quad \textbf{(8-12)}$$

Thus only $R(0, \omega)$ need be investigated. This function of one variable can be differentiated. Its minima can then be located and checked.

When a rational function has j-axis poles, these can be separated by partial fractions. That is, for one such pair of poles

$$Z(s) = \frac{K}{s + jb} + \frac{K^*}{s - jb} + Z_1(s) \quad \textbf{(8-13)}$$

where $Z_1(s)$ has no poles in the right half plane or on the j-axis. In the general discussion above, before the present 1-port case was considered, it was shown that K, the residue of the j-axis pole, must be real and positive if $Z(s)$ is to be PR. When K in Equation (8-13) is real and positive,

$$\text{Re}[Z(j\omega)] = \text{Re}[Z_1(j\omega)] \tag{8-14}$$

Furthermore, for $\text{Re}[s] > 0, \text{Re}[Z(s)] > \text{Re}[Z_1(s)]$. Thus, the minimum of $\text{Re}[Z(s)]$ is equal to the minimum of $\text{Re}[Z_1(s)]$ for $\text{Re}[s] \geqslant 0$.

In summary the above results are as follows: A rational function is a PR function if and only if

1. It has no right half plane poles.
2. Poles on the j-axis are simple and have positive real residues.
3. The real part is non-negative on the j-axis.

These three conditions constitute a readily applied criterion for a PR function.

***PR* 2-ports.** For a 2-port the quadratic form of the PR definition is

$$P_0(s) = H_{11}(s)\ |V_1|^2 + H_{12}\ V_1^*\ V_2 + H_{21}V_2^*\ V_1 + H_{22}|V_2|^2 \tag{8-15}$$

Since $P_0(s)$ must be PR for all V_1 and V_2, $H_{11}(s)$ and $H_{22}(s)$ must both be PR. $H_{12}(s)$ and $H_{21}(s)$ cannot have any right half plane poles, but they can have right half plane zeros. If $H_{12}(s)$ or $H_{21}(s)$ have j-axis poles, the same pole must appear in both $H_{11}(s)$ and $H_{22}(s)$. For example, if all the H's have a pole at $s = j\omega_0$, then near $s = j\omega_0$

$$P_0(s) \approx \frac{K_{11}|V_1|^2 + K_{12}V_1^*\ V_2 + K_{21}V_1V_2^* + K_{22}|V_2|^2}{(s - j\omega_0)} \tag{8-16}$$

where K_{ij} is the residue in the pole of $H_{ij}(s)$ at $s = j\omega_0$. The numerator in Equation (8-16), which is the residue in the pole of $P_0(s)$, must be real and positive for all complex V_1 and V_2. Both K_{11} and K_{22} must be real and positive; for if either V_1 or V_2 is zero, the numerator reduces to K_{22} or K_{11}, respectively, times a real positive number.

To get conditions on K_{12} and K_{21}, we separate the residue K into its real and imaginary parts.

$$\begin{aligned} K = K_{11}|V_1|^2 + K_{22}|V_2|^2 &+ \{\text{Re}[K_{12}] + \text{Re}[K_{21}]\}\ \text{Re}[V_1V_2^*] \\ &- \{\text{Im}[K_{12}] - \text{Im}[K_{21}]\}\ \text{Im}[V_1V_2^*] \\ &+ j\{[\text{Im}[K_{12}] + \text{Im}[K_{21}]]\ \text{Re}[V_1V_2^*] \\ &+ [\text{Re}[K_{12}] - \text{Re}[K_{21}]]\ \text{Im}[V_1V_2^*]\} \end{aligned} \tag{8-17}$$

When $K_{12} = K_{21}$, as in the case of the impedance or admittance matrix of a bilateral circuit, then K_{12} must be real. Furthermore, since

$$\text{Re}[V_1 V_2^*] \leqslant |V_1|\,|V_2|$$

the most stringent condition on K_{12} occurs when

$$K = K_{11}|V_1|^2 + K_{22}|V_2|^2 + 2K_{12}|V_1||V_2| \tag{8-18}$$

But this is exactly the case where the K_{ij} are the elements of a symmetric matrix used to construct a real quadratic form. Thus when $K_{12} = K_{21}$, the *PR* condition reduces to

$$K_{11} \geqslant 0;\ K_{22} \geqslant 0,\ K_{11}K_{22} - K_{12}^2 \geqslant 0$$

When K_{12} and K_{21} are both complex, the fact that $\text{Im}[K] = 0$ requires $K_{12} = K_{21}^*$, then Equation (8-17) becomes

$$K = K_{11}|V_1|^2 + K_{22}|V_2|^2 + 2\ \text{Re}\ [K_{12}\ V_1 V_2^*] \tag{8-19}$$

For any fixed $|V_1|$ and $|V_2|$, the angles can be selected so that $K_{12}\ V_1 V_2^*$ is real and negative. This is the worst condition. Again (8-19) reduces to the real quadratic form with K_{12} as the off-diagonal term of the corresponding symmetric matrix. Thus, in general, a necessary condition for *PR*ness is

$$\begin{array}{lll} K_{11}\ \text{real} & K_{22}\ \text{real} & K_{12} = K_{21}^* \\ K_{11} \geqslant 0 & K_{22} \geqslant 0 & K_{11}\ K_{22} - |K_{12}|^2 \geqslant 0 \end{array} \tag{8-20}$$

The above discussion of residues applies to n-ports when the ports are taken two at a time. The result corresponding to Equation (8-20) is

$$\begin{array}{l} K_{ii}\ \text{real and positive} \\ K_{ij} = K_{ji}^* \\ K_{ii}K_{jj} - |K_{ij}|^2 \geqslant 0 \end{array} \tag{8-21}$$

for all i and j, $i \neq j$.

In this section we have derived a straightforward criterion for testing rational functions to see if they are *PR*. Since the diagonal elements of a *PR* matrix must be a *PR* function, this procedure is an important step in testing matrixes of rational functions to see if they are *PR*. We also derived some necessary conditions on the location of the poles and zeros of all elements of a rational *PR* matrix. For j-axis poles we derived additional conditions that the residues must satisfy.

The next step should be the derivation of a complete procedure for testing matrixes to see if they are *PR*. Such a procedure is beyond the scope of this text. When a given matrix is symmetric, there are straightforward ways of testing for *PR*ness (Reference 15). The general procedures for other matrixes require the theory of Hermitian forms (Reference 9, Chapter 10). For 2-port and simple 3-port cases, the definition of *PR* plus the 1-port test is adequate.

Now that we know how to test a matrix to see if it is *PR* we might again ask "so what." The reason, as shown in the remainder of the chapter, is that *PR* systems have some interesting engineering properties. To the network theorist, symmetric *PR* matrixes are even more significant because it is known (Reference 16) that any such matrix can be realized as an n-port network of positive R's, L's, C's, and ideal transformers. The given *PR* matrix is then the admittance or impedance matrix of the network.

8-2 Causality of *PR* Systems and Relations between Real and Imaginary Parts

In Section 7-2 we pointed out that when a rational frequency-domain function was analytic in the right half plane, its real and imaginary parts for $s = j\omega$ were related by the Hilbert transform. Then the corresponding time function was zero for $t < 0$. When the transfer function matrix is *PR*, all the response functions are zero for $t < 0$ and the system is causal.

For an element of a *PR* matrix, or any transfer function derived from a causal system response function, the real part or imaginary part for $s = j\omega$ is sufficient to determine the complete function for all s. For example, if the real part, including possible impulses where the complete function has j-axis poles, is given, then the imaginary part can be constructed via the Hilbert transform. Finally, the substitution $\omega = -js$ gives the functional form for all s.

When a transfer function is rational and has no poles in the right half plane, then there is an easier way to generate the complete function than going through the Hilbert transform integrals. The real part is a rational function in ω^2. The complete response is readily determined from a partial fraction expansion of the real part.

Since

$$H(j\omega) = H^*(-j\omega) \qquad \text{Re}[H(j\omega)] = \tfrac{1}{2}[H(j\omega) + H(-j\omega)] \tag{8-22}$$

The derivation of the procedure begins by separating $H(s)$ into its even and odd parts. Thus,

$$H(s) = \frac{m_1 + n_1}{m_2 + n_2} \tag{8-23}$$

where

$$m_1 = A_0 + A_2 s^2 + \cdots + A_{2k} s^{2k}$$

$$n_1 = A_1 2 + A_3 s^3 + \cdots + A_{2j+1} s^{2j+1}$$

$$m_2 = B_0 + B_2 s^2 + \cdots + B_{2m} s^{2m}$$

$$n_2 = B_1 s + B_3 s^3 + \cdots + B_{2m+1} s^{2m+1}$$

The fact that m_2 and n_2 can differ in degree by at most one is a consequence of the *PR*ness of $[H_{ij}(s)]$.

Now

$$\text{Re } [H(j\omega)] = \frac{m_1(j\omega)\, m_2(j\omega) - n_1(j\omega)\, n_2(j\omega)}{m_2^2(j\omega) - n_2^2(j\omega)} \tag{8-24}$$

and

$$Ev[H(s)] = \frac{m_1 m_2 - n_1 n_2}{m_2^2 - n_2^2} = \tfrac{1}{2}[H(s) + H(-s)] \tag{8-25}$$

Since we assume no j-axis poles, $H(s)$ has only left half plane poles and $H(-s)$ has only right half plane poles. Thus, $Ev[H(s)]$ has poles with quadrantal symmetry.[2] The left half plane ones are the poles of $H(s)$. The constant multiplier of $H(s)$ is the same as that of $Ev[H(s)]$ because of the $\frac{1}{2}$ factor in the expression.

The construction of $H(s)$ from a given real part proceeds as follows: The given real part is a rational function in ω^2. We can construct $Ev[H(s)]$ by substituting $-s^2$ for ω^2. Then we can expand $Ev[H(s)]$ in partial fractions and throw away all right half plane terms. Finally we fix up the constant multiplier of the resulting function by comparing it to $Ev[H(0)]$. The above construction procedure, which is due to H. Bode, is extremely useful in certain filter problems.

As an example let us construct a function from a given real part. Given

$$\text{Re } [H(j\omega)] = \frac{1}{\omega^4 + 1} \tag{8-26}$$

then

$$Ev[H(s)] = \frac{1}{s^4+1} = \frac{\frac{1}{4\sqrt{2}}(1+j)}{s + \frac{1}{\sqrt{2}} + j\frac{1}{\sqrt{2}}} + \frac{\frac{1}{4\sqrt{2}}(1-j)}{s + \frac{1}{\sqrt{2}} - j\frac{1}{\sqrt{2}}}$$
$$+ \frac{\frac{1}{4\sqrt{2}}(-1-j)}{s - \frac{1}{\sqrt{2}} - j\frac{1}{\sqrt{2}}} + \frac{\frac{1}{4\sqrt{2}}(-1+j)}{s - \frac{1}{\sqrt{2}} + j\frac{1}{\sqrt{2}}} \tag{8-27}$$

[2] Quadrantal symmetry means four points located symmetrically about both real and imaginary axes. For the complex number α, the other three points are at $-\alpha$, α^*, and $-\alpha^*$.

The first two terms in the partial fraction expansion (8-27) are from $H(s)$ and the second two from $H(-s)$. Thus,

$$\tfrac{1}{2}H(s) = \frac{\frac{1}{4\sqrt{2}}(1+j)}{s+\frac{1}{\sqrt{2}}+j\frac{1}{\sqrt{2}}} + \frac{\frac{1}{4\sqrt{2}}(1-j)}{s+\frac{1}{\sqrt{2}}-j\frac{1}{\sqrt{2}}} = \frac{\frac{1}{2\sqrt{2}}s+\frac{1}{2}}{s^2+\sqrt{2}s+1} \tag{8-28}$$

or

$$H(s) = \frac{1}{\sqrt{2}}\left(\frac{s+\sqrt{2}}{s^2+\sqrt{2}\,s+1}\right) \tag{8-29}$$

Since $\mathrm{Re}[H(j\omega)]$ is positive and $H(s)$ has no right half plane or j-axis poles, $H(s)$ is *PR*.

As a second example let us consider a transfer function that is not *PR*. Let

$$\mathrm{Re}[H(j\omega)] = \frac{1-\omega^2}{\omega^4+1} \tag{8-30}$$

Then

$$\begin{aligned} Ev[H(s)] &= \frac{s^2+1}{s^4+1} \\ &= \frac{j(\sqrt{2}/4)}{s+\frac{1}{\sqrt{2}}+j\frac{1}{\sqrt{2}}} - \frac{j(\sqrt{2}/4)}{s+\frac{1}{\sqrt{2}}-j\frac{1}{\sqrt{2}}} + \tfrac{1}{2}H(-s) \end{aligned} \tag{8-31}$$

Thus,

$$H(s) = \frac{j(\sqrt{2}/2)}{s+\frac{1}{\sqrt{2}}+j\frac{1}{\sqrt{2}}} - \frac{j(\sqrt{2}/2)}{s+\frac{1}{\sqrt{2}}-j\frac{1}{\sqrt{2}}} = \frac{1}{s^2+\sqrt{2}\,s+1} \tag{8-32}$$

The partial fraction method works for any transfer function that is rational and analytic in the right half plane. Such a transfer function represents a causal, lumped, linear, time-invariant system.

When there are j-axis poles with real residues, then the real part contribution from these poles is zero except for impulses at the pole locations. The Hilbert transform can be used to get the imaginary part of the complete function due to these poles. When the transfer function has a j-axis pole with an imaginary residue, the pole appears in the real part. The imaginary part has an impulse at this pole location. For poles with complex residues, the situation can be handled as one real residue pole plus one imaginary residue pole.

Magnitude and phase relations. For driving point functions, the diagonal elements of the hybrid matrixes, the real part for $s = j\omega$ is an important quantity. It is a measure of the power delivered to the system by a real frequency source. For transfer functions the real part has little physical significance. The magnitude and phase of transfer functions are the important quantities. The quantities are related to the transfer function by

$$H(j\omega) = e^{-(\alpha(\omega)+j\beta(\omega))} = e^{\gamma(\omega)} \tag{8-33}$$

where $\alpha(\omega)$ is the loss (attenuation) function; $\beta(\omega)$ is the phase lag; $\gamma(\omega)$ is the propagation constant.

As a function of a complex variable

$$\gamma(-js) = \ln H(s) \tag{8-34}$$

Since $H(s)$ is analytic in the right half plane, $\gamma(-js)$ is analytic if $H(s)$ is nonzero. Of course if $H(s)$ has j-axis poles, $\gamma(\omega)$ is not defined at the pole locations. Thus, if a transfer function has no right half plane zeros and no poles or zeros on the j-axis, then $\gamma(s)$ is analytic in the right half plane and on the boundary of that plane. Along this boundary (the j-axis) the Hilbert transform relates the real and imaginary parts. In other words, $\alpha(\omega)$ and $\beta(\omega)$ are related by the Hilbert transform. Since α is even and β is odd, the forms (7-42) and (7-43) of the transform apply as well as (7-40) and (7-41).

A transfer function with no right half plane zeros is called a *minimum phase* transfer function. The reason for the term is that for any non-minimum phase transfer function, there is another transfer function with the same magnitude and smaller phase angle. To see this suppose $H(s)$ has a pair of conjugate right half plane zeros at s_1 and $s_1{}^*$. The factors $(s - s_1)(s - s_1{}^*)$ appear in the numerator of $H(s)$. To change these to factors $(s + s_1)(s + s_1{}^*)$ we merely multiply $H(s)$ by

$$\frac{(s + s_1)(s + s_1{}^*)}{(s - s_1)(s - s_1{}^*)}$$

The magnitude of this factor is one for real frequencies. Thus $\alpha(\omega)$ does not change when $H(s)$ is multiplied by the factor. From an s-plane pole zero sketch, the phase angle of such a factor is seen to be negative for all positive frequencies. Thus the factor increases the phase lag. When all the zeros are moved over to the left half plane, the resulting $H(s)$ is the unique minimum phase function with the given attenuation.

Although there is a unique minimum phase transfer function corresponding to any given attenuation, there are infinitely many nonminimum phase ones. If s_1 is in the left half plane, then

$$H_1(s) = \frac{(s - s_1)(s - s_1{}^*)}{(s + s_1)(s + s_1{}^*)} \tag{8-35}$$

is a transfer function with magnitude one and a positive phase lag. Multiplication of any transfer function by such a factor increases the lag, but does not disturb the attenuation. Factors of this sort are called *all pass* factors. A lattice with a reactance function in the series arms and the dual in the shunt arms is an all pass circuit. Figure 8-1 is an example of such a circuit.

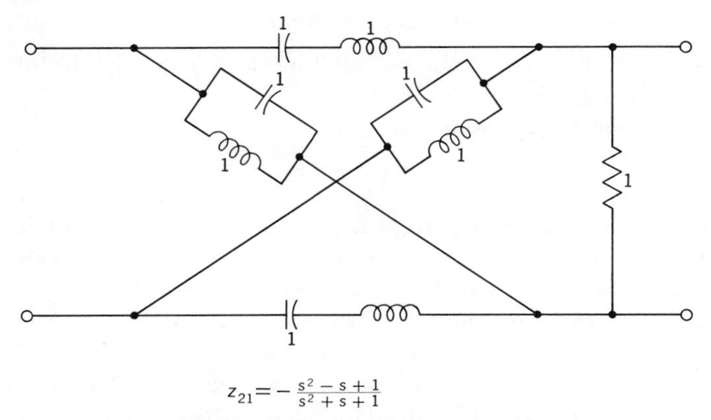

$$z_{21} = -\frac{s^2 - s + 1}{s^2 + s + 1}$$

Figure 8-1 An all pass circuit.

Another pair of quantities associated with transfer functions that have considerable physical significance are the derivative of the loss function, known as the *logarithmic decrement*, and the derivative of the phase lag, known as the *group delay*. When

$$\begin{aligned} H(j\omega) &= R(\omega) + jX(\omega) \\ \alpha(\omega) &= -\ln \sqrt{R^2(\omega) + X^2(\omega)} \end{aligned} \tag{8-36}$$

and

$$\beta(\omega) = -\tan^{-1} \frac{X(\omega)}{R(\omega)} \tag{8-37}$$

Then the logarithmic decrement is

$$L(\omega) = \frac{R(\omega)\, R'(\omega) + X(\omega)\, X'(\omega)}{R^2(\omega) + X^2(\omega)} \tag{8-38}$$

where the prime indicates differentiation with respect to ω. Similarly the group delay is

$$T_d(\omega) = \frac{X'(\omega)\, R(\omega) - R'(\omega)\, X(\omega)}{R^2(\omega) + X^2(\omega)} \tag{8-39}$$

Both $L(\omega)$ and $T_d(\omega)$ are rational. The logarithmic decrement is an odd function and the group delay is even. Since the derivative of an analytic function is analytic, the function

$$\gamma'(s) = T_d(-js) + jL(-js) \tag{8-40}$$

is analytic in the right half plane when the transfer function is minimum phase. Thus for minimum phase transfer functions, T_d and L are related by the Hilbert transform. Furthermore, since they are both rational, the Bode procedure can be used to find $\gamma'(s)$ from a given $T_d(\omega)$.

Some examples. To show the relations between α, β, L, and T_d more explicitly let us return to the example whose real part is (8-30) and whose complete transfer function is (8-32). For this example

$$H(j\omega) = \frac{1-\omega^2}{\omega^4+1} - j\frac{\sqrt{2}\,\omega}{\omega^4+1} \tag{8-41}$$

$$|H(j\omega)|^2 = \frac{1}{\omega^4+1} \tag{8-42}$$

$$\alpha(\omega) = \tfrac{1}{2}\ln\,(\omega^4+1) \tag{8-43}$$

$$\beta(\omega) = -\tan^{-1}\frac{(-\sqrt{2}\,\omega)}{1-\omega^2} \tag{8-44}$$

$$T_d(\omega) = \frac{\sqrt{2}\,(\omega^2+1)}{\omega^4+1} \tag{8-45}$$

$$L(\omega) = \frac{2\,\omega^3}{\omega^4+1} \tag{8-46}$$

We can verify that these last two quantities are related as described above by applying the Bode procedure to (8-45) and obtaining (8-46).

$$\begin{aligned} Ev(\gamma'(s)) &= \frac{\sqrt{2}\,(1-s^2)}{s^4+1} \\ &= \frac{\frac{1}{2}}{s+\frac{1}{\sqrt{2}}+j\frac{1}{\sqrt{2}}} + \frac{\frac{1}{2}}{s+\frac{1}{\sqrt{2}}-j\frac{1}{\sqrt{2}}} + \tfrac{1}{2}\,\gamma'(-s) \end{aligned} \tag{8-47}$$

Thus,

$$\gamma'(s) = \frac{2\,s+\sqrt{2}}{s^2+\sqrt{2}\,s+1} \tag{8-48}$$

$$\gamma'(j\omega) = \frac{\sqrt{2}\,(\omega^2+1)}{\omega^4+1} + \frac{2\,\omega^3}{\omega^4+1} \tag{8-49}$$

As expected, Equation (8-49) checks with Equations (8-45) and (8-46).

As a second example let us consider the transfer function (8-29). This function has a left half plane zero at $s = -\sqrt{2}$. If we multiply this transfer function by the factor $(s-\sqrt{2})/(s+\sqrt{2})$, a factor whose magnitude is one, we get a second transfer function that is not minimum phase. Let us examine the various quantities relative to these two transfer functions. Let

$$H_1(s) = \frac{1}{\sqrt{2}} \frac{s + \sqrt{2}}{s^2 + \sqrt{2}\,s + 1} \tag{8-50}$$

$$H_2(s) = \frac{1}{\sqrt{2}} \frac{s - \sqrt{2}}{(s^2 + \sqrt{2}\,s + 1)} \tag{8-51}$$

$$H_1(j\omega) = \frac{1}{\omega^4 + 1} - j\,\frac{\omega(\omega^2 + 1)}{\sqrt{2}\,(\omega^4 + 1)} \tag{8-52}$$

$$H_2(j\omega) = \frac{2\omega^2 - 1}{\omega^4 + 1} - j\,\frac{\omega(\omega^2 - 3)}{\sqrt{2}\,(\omega^4 + 1)} \tag{8-53}$$

$$|H_1(j\omega)|^2 = \frac{\omega^2 + 2}{2(\omega^4 + 1)} \tag{8-54}$$

$$|H_2(j\omega)|^2 = \frac{\omega^2 + 2}{2(\omega^4 + 1)} \tag{8-55}$$

$$\alpha_1(\omega) = \tfrac{1}{2}\,[\ln 2 + \ln\,(\omega^4 + 1) - \ln\,(\omega^2 + 2)] \tag{8-56}$$

$$\alpha_2(\omega) = \tfrac{1}{2}\,[\ln 2 + \ln\,(\omega^4 + 1) - \ln\,(\omega^2 + 2)] \tag{8-57}$$

$$\beta_1(\omega) = -\tan^{-1} - \omega(\omega^2 + 1) \tag{8-58}$$

$$\beta_2(\omega) = -\tan^{-1} \frac{-\omega(\omega^2 - 3)}{2\omega^2 - 1} \tag{8-59}$$

For the first network the phase shift, which is the negative of the phase lag $\beta_1(\omega)$, starts at zero when ω is zero, and decreases monotonically as ω increases. The phase shift $-\beta_2(\omega)$ starts at π when $\omega \approx 0$ and decreases monotonically. At all times $-\beta_2 > -\beta_1$. Thus, the phase shift of the minimum phase system, the first system, is less than that of the second.

As an exercise the student can compute the group delay and the logarithmic decrement for the two systems. He should then apply the Bode procedure to both group delay functions and see that only the first gives the right logarithmic decrement.

Asymptotic limits on magnitude and phase. Often in system design problems, a desired property for a system is a high-frequency attenuation that is essentially infinite. Such ideal filtering characteristics are not possible with lumped, linear, time-invariant systems. Through the Hilbert transform relations we can derive a maximum rate of increase of attenuation with frequency for high frequencies. That is, we can get an asymptotic bound on the attenuation. This bound is known as the *Paley-Wiener criterion.* A complete proof is given in Reference 21, Theorem XII, pp. 16–17.

The basic idea behind the Paley-Wiener criterion is that for *PR* systems, the real and imaginary parts of all transfer functions are well-defined and related by the Hilbert transform. The most convenient starting point to derive the specific result is formula (7-42). In terms of the logarithm of the magnitude $\alpha(\omega)$ and phase lag $\beta(\omega)$, that formula is

$$\beta(\omega) = \frac{2\omega}{\pi} \int_0^\infty \frac{\alpha(\lambda)}{\omega^2 - \lambda^2} \, d\lambda \tag{8-60}$$

If α increases too rapidly with increasing ω, then the integral will not exist. If for large ω, $\alpha(\omega)$ increases as the first power of ω, the integral increases logarithmically as the upper limit goes to infinity. Thus, the integral with the infinite limit does not exist. If the rate of growth of $\alpha(\omega)$ is less than the first power of ω, then the integral does exist. Since the assumption of a well-defined magnitude and phase relation would be false if the integral did not exist, we conclude that $\alpha(\omega)$ cannot grow as fast as ω.

The Paley-Wiener criterion is the translation of the above result to a condition on the transfer function $H(j\omega)$. It is:

For a *PR* system a transfer function cannot approach zero as fast as an exponential as $\omega \to \infty$. In symbols,

$$\lim_{\omega \to \infty} \frac{Ke^{-\epsilon\omega}}{|H(j\omega)|} = 0$$

for all real positive K and ϵ.

In the lumped system case $H(j\omega)$ is rational. Thus as $\omega \to \infty$, $|H(j\omega)| \to 1/\omega^n$ for integer n is the best that can be done. Such a transfer function is allowed by the Paley-Wiener condition. The condition shows that no other causal system can do better.

A corresponding condition on the phase can be derived starting with formula (7-43). In α, β terms that is

$$\alpha(\omega) = \frac{2}{\pi} \int_0^\infty \frac{\lambda\beta(\lambda)}{\omega^2 - \lambda^2} \, d\lambda \tag{8-61}$$

From this formula it appears that $\beta(\omega)$ must go to zero as $\omega \to \infty$. However, we know that in most circuits the phase lag increases to some constant value. The difficulty lies in the fact that $\alpha(\omega)$ is not uniquely determined by the Hilbert transform even for a minimum phase transfer function. There is always the possibility of an additive constant.

If we look at $\alpha(\omega) - \alpha(0)$ we get rid of the constant

$$\begin{aligned} \alpha(\omega) - \alpha(0) &= \frac{2}{\pi} \int_0^\infty \left[\frac{\lambda\beta(\lambda)}{\omega^2 - \lambda^2} - \frac{\lambda\beta(\lambda)}{-\lambda^2} \right] d\lambda \\ &= \frac{2}{\pi} \int_0^\infty \frac{\omega^2\beta(\lambda)}{\lambda(\omega^2 - \lambda^2)} \, d\lambda \end{aligned} \tag{8-62}$$

From this last formula we see that so long as $\beta(\lambda)$ does not increase as fast as λ^2, the integral exists. Thus, not only is an asymptotically constant phase possible, but a phase lag that grows linearly or something like $\omega^{3/2}$ is possible in the general *PR* system.

8-3 Stability for *PR* Systems

Stability is an intuitive concept that requires a fair bit of qualification when one wants to make a precise definition. For linear systems there are at least three different types of stability that are significant. In words, these three are:

1. *Input-output stability:* Every bounded input gives a bounded output.
2. *Asymptotic stability:* For any finite set of initial conditions (well-defined state at t_0) and zero input for $t > t_0$, all states approach zero asymptotically as t approaches infinity.
3. *Bounded stability:* For any finite set of initial conditions at t_0 and zero input for $t > t_0$, all states remain bounded for t greater than t_0.

When a lumped, linear, time-invariant system is known to be characterized by a *PR* matrix, its input-output stability is determined. If the system is controllable and observable it is bounded stable. If no element of the transfer function matrix has a j-axis pole, including the point at infinity, then the system is input-output stable. If the system is input-output stable, controllable and observable, it is also asymptotically stable. The bounded stability and asymptotic stability are easy to show and therefore are left to the problems. The input-output stability requires a more involved proof as given below.

The key step in showing the conditions for input-output stability is the following well-known theorem [see Reference 28, Section 8.4].

Theorem: Any single-input–single-output system characterized by a response function $h(t)$ and the input-output relation

$$y(t) = \int_{-\infty}^{\infty} h(t-\tau)\, u(\tau)\, d\tau \tag{8-63}$$

is input-output stable if and only if $\int_{-\infty}^{\infty} |h(t)|\, dt$ exists.

Proof: To prove the *if* statement we first define[3]

[3] Since the hypothesis is $u(t)$ and is bounded, there is a least upper bound N. If $u(t)$ is not continuous it may not really have a maximum value. In this case we should use sup. for supremum, or l.u.b. instead of max.

$$M = \int_{-\infty}^{\infty} |h(t)|\, dt \tag{8-64}$$

$$N = \max |u(t)| \tag{8-65}$$

Now by a well-known property of integrals

$$\begin{aligned} |y(t)| &= \left| \int_{-\infty}^{\infty} h(t-\tau)\, u(\tau)\, d\tau \right| \\ &\leqslant \int_{-\infty}^{\infty} |h(t-\tau)||u(\tau)|\, d\tau \\ &\leqslant N \int_{-\infty}^{\infty} |h(t-\tau)|\, d\tau = N \int_{-\infty}^{\infty} |h(\nu)|\, d\nu \\ &= NM \end{aligned} \tag{8-66}$$

Thus $|y(t)|$ is bounded by NM; therefore, the *if* part of the theorem is proven.

To show the *only if* statement, we assume $\int_{-\infty}^{\infty} |h(t)|\, dt$ does not exist and show that there is a bounded input such that $y(t)$ does not exist for some t. To make the notation simple we pick $t = 0$ and $u(t) = \text{sgn}\,(h(-t))$.[4] Now

$$y(t) = \int_{-\infty}^{\infty} h(t-\tau)\ \text{sgn}\,(h(-\tau))\, d\tau \tag{8-67}$$

Then

$$\begin{aligned} y(0) &= \int_{-\infty}^{\infty} h(-\tau)\ \text{sgn}\,(h(-\tau))\, d\tau \\ &= \int_{-\infty}^{\infty} |h(-\tau)|\, d\tau = \int_{-\infty}^{\infty} |h(\nu)|\, d\nu \end{aligned} \tag{8-68}$$

Since the integral does not exist, $y(0)$ is not bounded and the theorem is proven.

The extension of the theorem to systems with several inputs and several outputs is obvious since each output is a finite sum of convolutions. Since the response functions of a system characterized by a *PR* matrix with no j-axis poles are impulses at $t = 0$ plus damped exponentials starting at $t = 0$, all response functions are absolutely integrable.

---

[4] The function sgn (X) is defined by

$$\text{sgn}\,(X) = \begin{cases} -1 & \text{if } X < 0 \\ 0 & \text{if } X = 0 \\ 1 & \text{if } X > 0 \end{cases}$$

8-4 Passivity for *PR* Systems

From the definition of a *PR* matrix and the properties of transforms in Section 8-2 it is easy to show that a system characterized by a *PR* matrix is passive. To demonstrate passivity we must examine the power delivered to the network. In our notation, the power $p(t)$ is

$$p(t) = \mathbf{v}^T(t)\ \mathbf{y}(t) \tag{8-69}$$

Since the power is the sum of products of time functions, its Laplace-Fourier transform is the convolution as in Equation (7-35). Thus,

$$P(j\omega) = \frac{1}{2\pi} \int_{-\infty}^{\infty} \mathbf{V}^T(j(\omega - \lambda))\ \mathbf{Y}(j\lambda)\ d\lambda \tag{8-70}$$

But by the definition of the transfer function matrix

$$\mathbf{Y}(j\omega) = \underset{\sim}{H}(j\omega)\ \mathbf{V}(j\omega) \tag{8-71}$$

Thus

$$P(j\omega) = \frac{1}{2\pi} \int_{-\infty}^{\infty} \mathbf{V}^T(j(\omega - \lambda))\ \underset{\sim}{H}(j\lambda)\ \mathbf{V}(j\lambda)\ d\lambda \tag{8-72}$$

Since systems characterized by *PR* matrixes are causal, the power depends only on values of the excitation for past times. Thus, if we define an excitation

$$\mathbf{v}_1(t) = \begin{cases} \mathbf{v}(t) & \text{for } t \leq t_1 \\ 0 & \text{for } t > t_1 \end{cases} \tag{8-73}$$

then $p_1(t)$ is

$$p_1(t) = \begin{cases} p(t) & \text{for } t \leq t_1 \\ 0 & \text{for } t > t_1 \end{cases} \tag{8-74}$$

In addition

$$\int_{-\infty}^{t_1} p(\tau)\ d\tau = \int_{-\infty}^{\infty} p_1(\tau)\ d\tau \tag{8-75}$$

The derivation of Equation (8-72) applies to $P_1(j\omega)$ just as it does to $P(j\omega)$. Furthermore,

$$P_1(0) = \int_{-\infty}^{\infty} p_1(t)\ dt = \int_{-\infty}^{t_1} p(\tau)\ d\tau \tag{8-76}$$

Thus, if we can show that $P_1(0)$ is positive for all excitation vectors $\mathbf{v}(t)$ independent of the time t_1, then we have proven that the system is passive. From Equation (8-72)

$$P_1(0) = \frac{1}{2\pi} \int_{-\infty}^{\infty} \mathbf{V}_1{}^T(-j\lambda)\ \underset{\sim}{H}(j\lambda)\ \mathbf{V}_1(j\lambda)\ d\lambda \tag{8-77}$$

But since $\mathbf{v}_1(t)$ is a vector of real time functions,

$$\mathbf{V}_1(-j\lambda) = \mathbf{V}_1^*\ (j\lambda) \tag{8-78}$$

Thus,

$$P_1(0) = \frac{1}{2\pi} \int_{-\infty}^{\infty} \mathbf{V}_1^{T*}\ (j\lambda)\ \underset{\sim}{H}(j\lambda)\ \mathbf{V}_1(j\lambda)\ d\lambda \tag{8-79}$$

By the definition of the *PR*ness of $H(s)$, the integrand is positive for all λ. Thus, $P_1(0)$ is positive regardless of the exact nature of $\mathbf{v}_1(t)$. So long as $\mathbf{V}_1(j\omega)$ exists, $P_1(0)$ is positive. It can be shown that any signal that can deliver only a finite amount of power has a well-defined transform for $s = j\omega$. Thus, a system characterized by a *PR* matrix is passive.

An example. So far we have seen that a system with positive semidefinite energy functions W_C, W_L, and P_d is passive and its hybrid transfer function matrix is *PR*. Furthermore, we have shown that any system characterized by a *PR* matrix is passive. It can also be shown that every passive system has a hybrid matrix that is *PR*, but this requires more complex variable theory than that assumed as background for this text (see Reference 18). For the class of systems that can be modeled with controlled sources as well as *R*'s, *L*'s, and *C*'s, the energy functions need not all be positive semidefinite for passivity or *PR*ness. The following 1-port circuit demonstrates this last point.

Consider the circuit of Figure 8-2. Find the range of values for g_m

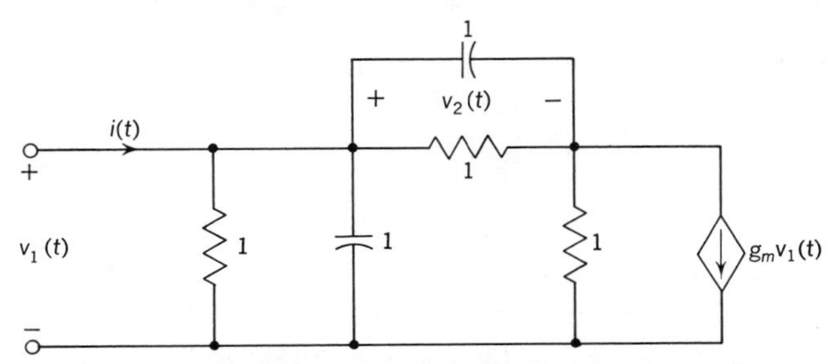

Figure 8-2 An example that is *PR* but not *PSD*.

such that (a) the input impedance is *PR*, and (b) the dissipative energy function P_d is positive semidefinite.

The normal form equations are

$$\begin{aligned} \dot{v}_1 &= (-2 - g_m)\ v_1 + v_2 + i(t) \\ \dot{v}_2 &= (1 + g_m)\ v_1 - 2v_2 \end{aligned} \tag{8-80}$$

The energy and dissipation functions are

$$\begin{aligned} W_L &= 0 \qquad W_C = \tfrac{1}{2}\ [v_1^2 + v_2^2] \\ P_d &= (2 + g_m)\ v_1^2 - (2 + g_m)\ v_1 v_2 + 2v_2^2 \end{aligned} \tag{8-81}$$

The conditions for positive semidefiniteness of P_d are

$$\begin{aligned} 2 + g_m &\geq 0 \\ 2(2 + g_m) - (1 + \tfrac{1}{2} g_m)^2 &\geq 0 \end{aligned} \tag{8-82}$$

The second inequality requires

$$-2 \leq g_m \leq 6 \tag{8-83}$$

This range for g_m also satisfies the first inequality. The criterion for the positive semidefiniteness of P_d is Equation (8-83).

The input impedance is readily found by transforming Equation (8-80) to the frequency domain and solving for V_1/I. Thus

$$\begin{aligned} j\omega\, V_1 &= (-2 - g_m)\, V_1 + V_2 + I \\ j\omega\, V_2 &= (1 + g_m)\, V_1 - 2V_2 \end{aligned} \tag{8-84}$$

The result is

$$Z(j\omega) = \frac{2 + j\omega}{(j\omega)^2 + (4 + g_m)(j\omega) + (3 + g_m)} \tag{8-85}$$

Now

$$\begin{aligned} &\text{Re}\,[Z(\sigma + j\omega)] \\ &= \frac{(2+\sigma)[\sigma^2 - \omega^2 + (4 + g_m)\,\sigma + 3 + g_m] + \omega^2(2\sigma + 4 + g_m)}{[\sigma^2 - \omega^2 + (4 + g_m)\,\sigma + 3 + g_m]^2 + \omega^2(2\sigma + 4 + g_m)^2} \end{aligned} \tag{8-86}$$

Since the denominator of this expression is a squared magnitude and thus always positive, the *PR*ness of $Z(s)$ depends only on the numerator. Thus, for $\sigma \geq 0$, we require

$$(2 + \sigma)\ [\sigma^2 + (4 + g_m)\ \sigma + 3 + g_m] + \omega^2\ (2 + \sigma + g_m) \geq 0 \tag{8-87}$$

Clearly this inequality is satisfied for all positive g_m. For $g_m < -2$, the coefficient of ω^2 is negative for $\sigma = 0$. Then for very large ω, the inequality can be violated for $g_m < -2$. This is the worst case.

In summary, we have $-2 \leq g_m \leq 6$ is required for the positive semidefiniteness of P_d, yet only $g_m > -2$ is required for *PR*ness and passivity.

PROBLEMS

8-1 For the following functions, construct the real part for Re $[s] > 0$ and see if each is *PR*.

a. $\dfrac{s + 2}{(s + 1)(s + 3)}$

b. $\dfrac{1}{s^2 + \sqrt{2}\, s + 1}$

c. $\dfrac{(s+1)^2}{(s+100)^2}$

8-2 Consider the following two matrixes as impedance parameter matrixes. Construct the Y matrixes and both hybrid matrixes and show that all satisfy the PR definition.

a. $\begin{bmatrix} s + \dfrac{1}{s} & \dfrac{1}{s} \\ \dfrac{1}{s} & \dfrac{1}{s} \end{bmatrix}$

b. $\begin{bmatrix} \dfrac{s + \sqrt{2}}{\sqrt{2}\, s^2 + 2s + \sqrt{2}} & \dfrac{1}{s^2 + \sqrt{2}\, s + 1} \\ \dfrac{1}{s^2 + \sqrt{2}\, s + 1} & \dfrac{s^2 + 1}{s^2 + \sqrt{2}\, s + 1} \end{bmatrix}$

8-3 For a 2-port with given impedance matrix show that the admittance matrix and both hybrid matrixes are PR if and only if the original impedance matrix is PR, provided that all matrixes exist except for poles.

8-4 **a.** Suppose Re $[H(j\omega)] = \dfrac{1}{1+\omega^8}$. Find $H(j\omega)$ that is a diagonal element of a PR matrix.

b. Suppose $|H(j\omega)|^2 = \dfrac{1}{1+\omega^8}$. Find $H(j\omega)$ that can be an off-diagonal element of a PR matrix.

8-5 Compute the phase lag $\beta(\omega)$ and the time delay $T_d(\omega)$ for the all pass factor (8-35).

8-6 Consider a transfer function

$$H(s) = \left(\frac{1}{s+a}\right)\left(\frac{(s - s_1)(s - s_1^*)}{(s + s_1)(s + s_1^*)}\right)$$

Choose the complex number s_1 with Re $[s_1] > 0$ so that

$$\int_0^a (T_d(j\omega) - T_d(0))^2 \, d\omega$$

is minimized.

8-7 Compute the group delay and logarithmic decrement for the transfer functions (8-50) and (8-51). Show that the two quantities are related by the Hilbert transform in the case of (8-50) but not in the case of (8-51).

8-8 For undistorted transmission of pulses it is desirable to have a system with a linear phase lag versus frequency; that is,

$$\beta(\omega) = \begin{bmatrix} K\omega & \text{for } |\omega| < \omega_0 \\ K\omega_0 & \text{for } |\omega| \geqslant \omega_0 \end{bmatrix}$$

Use the Hilbert transform to get an approximation to the associated magnitude. *Hint:* The Hilbert transform is a convolution integral. By differentiating $\beta(\omega)$ to impulses and integrating $1/\omega$ the same number of times, the convolution is easily performed. Appropriate choices for the constants of integration give

$$\int_{-\infty}^{\omega} \frac{-1}{\pi\lambda}\, d\lambda = \frac{-\ln \omega^2}{2\pi}$$

$$\int_{-\infty}^{\omega} \frac{-\ln \lambda^2}{2\pi}\, d\lambda = \frac{1}{2\pi}(\omega \ln \omega^2 - 2\omega)$$

8-9 Show that a controllable and observable system characterized by a *PR* matrix is bounded stable.

8-10 Show that for lumped, linear, time-invariant systems input-output stability implies asymptotic stability provided the system is both controllable and observable.

8-11 A lumped, linear, time-variant system is characterized by the zero-state response

$$y(t) = \int_{t_0}^{t} h(t, \tau)\, v(\tau)\, d\tau$$

Show that the necessary and sufficient condition for input-output stability is $\int_{-\infty}^{\infty} |h(t, \tau)|\, d\tau$ exists for all t.

8-12 For the circuit of Figure P8-12:
a. Consider v_1 as a source. Find the dissipative power P_d as a quadratic form of the state variables i_1 and i_2 and the parameter r_m. Find the range of r_m for which P_d is positive definite.
b. Compute the input impedance $Z(s)$ and find the range of r_m for which Re $[Z(j\omega)] \geqslant 0$ for all ω.

8-13 Consider the 2-port of Figure P8-13.
a. Consider i_1 and i_2 as sources. Find the dissipative power P_d as a

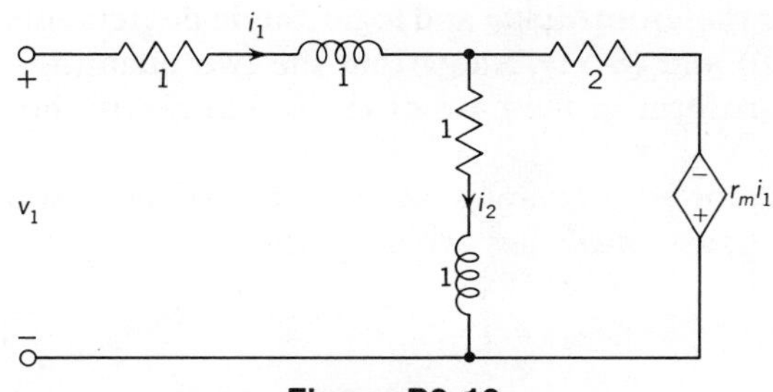

Figure P8-12

quadratic form in the state variables v_1, v_2 and possibly the inputs i_1 and i_2. Find the range of g_m for which P_d is positive definite.

b. Compute the admittance matrix $Y(s)$ and find the range of g_m for which this matrix is PR.

c. Compare the results of parts **a** and **b** with the example of Section 8-4.

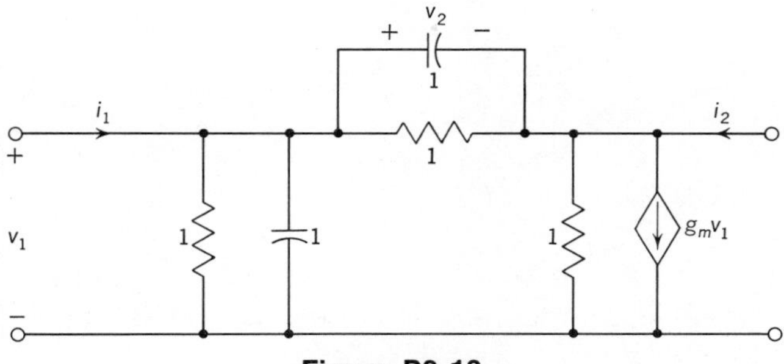

Figure P8-13

PART

III

Some Properties of and Techniques Applicable to Nonlinear Systems

CHAPTER

9

The Use of State Variable and Energy Function Concepts

In the class of physical systems that can be modeled by lumped system diagrams, the linear systems are special, and the linear, time-invariant systems are even more special. The reason that this last very special class is so important is that it is the only class of systems for which the general functional form of the signal processing operation is known in detail. Thus when a desired signal processing operation can be performed by a linear, time-invariant system, the engineer tries to use such a system. This is the only case wherein he can readily relate system components to signal processing requirements.

Two situations arise when it is either not possible or not desirable to use a linear, time-invariant system for a particular signal processing operation. The first is when the desired signal processing operation is not in the class that can be performed by a linear, time-invariant system. An example of this situation is the problem of an oscillator that must deliver amplitude stable a-c to a variable load with a fixed d-c source as input. The second situation arises when the desired operation is linear, but the actual physical devices used are not linear for the total range of allowed inputs. An example of this is an automobile suspension system that operates linearly over moderate bumps, but is driven into the non-linear range of the springs on very rough roads.

Some of the methods and concepts of Parts I and II of this text apply to nonlinear systems as well as linear. The concept of normal form differential equations formulated from an elemental diagram still applies. The use of block diagrams can be extended to include nonlinear elements. The concept of energy and dissipation functions is also useful in studying the stability of nonlinear systems.[1] This chapter is devoted to the application of these methods to nonlinear systems.

The next chapter introduces an additional analysis tool, the method of iteration. This technique, which can be applied to linear, time-variant and a fairly general class of nonlinear systems is useful both for analysis of nonlinear systems and as a method of proof of stability theorems.

9-1 Formulation

In the engineering literature there are two common mathematical forms for representing lumped, nonlinear systems. One is a system of first-order differential equations. The other is a system of integral equations, one equation for each nonlinear element. In this section the formulation of nonlinear problems in the two standard forms is considered. In the remainder of the chapter some of the more useful engineering applications of the differential equation forms are presented. The integral equation form is discussed in the next chapter. A detailed discussion of either approach would take an entire book.

For nonlinear systems we use the same two basic diagram forms that we used for linear systems: the circuit diagram and the block diagram. In linear, time-invariant systems the basic block is $\longrightarrow\boxed{h(t)}\longrightarrow$, which means $y(t) = h^*x(t)$. For nonlinear systems we add a second block: $\longrightarrow\boxed{f(x)}\longrightarrow$, which means $y(t) = f(x(t))$. Thus, the system of Figure 9-1 is characterized by

$$\left.\begin{aligned} v &= u - \dot{y} \\ x &= h*v \\ y &= f(x) \end{aligned}\right\} \quad x = h*u - h*f(x) \tag{9-1}$$

When written out, Equation (9-1) is seen to be the nonlinear integral equation

$$x(t) = \int_{-\infty}^{\infty} h(t-\tau)\, u(\tau)\, d\tau - \int_{-\infty}^{\infty} h(t-\tau)\, f(x(\tau))\, d\tau \tag{9-2}$$

[1] For nonlinear systems the concept of coenergy must be defined for a complete description on an energy basis. This concept is not needed for any of the developments here. Reference 5 has a discussion of energy and coenergy in nonlinear systems.

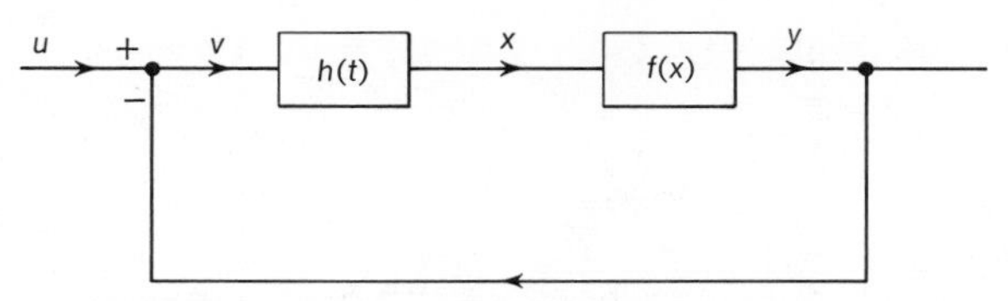

Figure 9-1 Basic nonlinear feedback loop.

If h is causal and $u(t)$ is zero for $t < t_0$, the integrals are proper and the equation takes the standard Volterra equation form in the mathematical literature. First we note that $\int_{t_0}^{t} h(t-\tau)\, u(\tau)\, d\tau$ is a known function of t; call it $g(t)$. Then Equation (9-2) becomes

$$x(t) = g(t) - \int_{t_0}^{t} h(t-\tau)\, f(x(\tau))\, d\tau \tag{9-3}$$

If $h(t)$ is a function in the usual sense; that is, no impulses or derivatives of impulses, then the Volterra theory (see Reference 27) applies. If h contains impulses or doublets, there are techniques for handling the equation. Some of these are discussed in the next chapter.

The differential equation characterization of the block diagram in Figure 9-1 is obtained by first describing the lumped, linear system in terms of state variables. If $h(t)$ contains no impulses, then the block characterized by $h(t)$ is also characterized by

$$\dot{\mathbf{w}} = \underset{\sim}{A}\, \mathbf{w} + \mathbf{B}\, v$$

$$x = \underset{\sim}{C}\, \mathbf{w}$$

Then, since $y = f(x)$ and $v = u - y$, the feedback system is characterized by the first-order equations

$$\dot{\mathbf{w}} = \underset{\sim}{A}\, \mathbf{w} - \mathbf{B}\, f(\underset{\sim}{C}\, \mathbf{w}) + \mathbf{B}\, u \tag{9-4}$$

The vector equation (9-4) can be written more compactly as

$$\dot{\mathbf{w}} = \mathbf{F}\, (\mathbf{w}, u) \tag{9-5}$$

where $\mathbf{F}$ is a vector-valued function of the $(n+1)$ vector $(w_1, w_2, \cdots, w_n, u)^T$. Before proceeding we should note that if $h(t)$ had contained an impulse or doublet, the vector equation (9-4) would not fall out so easily.

For electric circuit diagrams the nonlinear elements are often nonlinear two-terminal resistors, inductors, or capacitors. Nonlinear multiport elements can be handled in many cases, but even the simplest of these are beyond the scope of this text. A two-terminal resistor is described by a characteristic curve in its voltage-current plane. When the current is a function of the voltage, as in Figure 9-2(a), the resistor is said to be voltage-controlled. When the characteristic curve is that of Figure 9-2(b), where the voltage is a function of the current, the

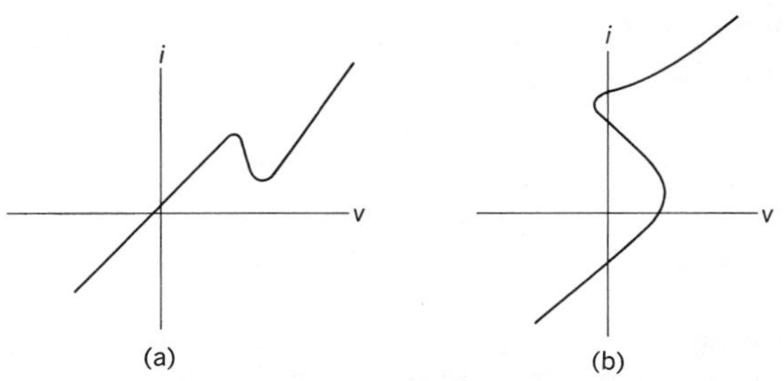

Figure 9-2 Nonlinear resistor characteristics. **(a)** A voltage-controlled resistor. **(b)** A current-controlled resistor.

resistor is current-controlled. When the characteristic curve is monotonic, then the resistor is both current- and voltage-controlled.

A nonlinear resistance is defined by its characteristic curve. Often instead of using the voltage-current relationship directly, it is convenient to use a formula that is more like a resistance or a conductance. For a current-controlled nonlinear resistor the voltage is a function of the current. Thus, the characteristic curve is defined by

$$v(t) = \bar{v}(i(t))$$

where $v(t)$ and $i(t)$ are the voltage and current as functions of time, and $\bar{v}$ represents the functional relation between voltage and current.

For such a resistor there are two different definitions of resistance that are used: the incremental resistance and the total resistance. The incremental resistance is the slope of the $v - i$ characteristic. The total resistance is that quantity which when multiplied by the square of current gives power. Thus if we use a prime to designate derivative with respect to an argument that is not time, the two resistances are

$$\begin{aligned} &\text{Incremental resistance:} \quad r(i) = \bar{v}'(i) \\ &\text{Total resistance:} \quad R(i) = \frac{\bar{v}(i)}{i} \end{aligned} \tag{9-6}$$

When a nonlinear resistor is voltage-controlled, a nonlinear conductance can be used to characterize the element. In this case the voltage-current characteristic is defined by

$$i(t) = \bar{i}(v(t)) \tag{9-7}$$

The incremental conductance is

$$g(v) = \bar{i}'(v) \tag{9-8}$$

The total conductance is

$$G(v) = \frac{\bar{i}(v)}{v} \tag{9-9}$$

For a monotonic resistor, both the resistances and conductances are well-defined. Furthermore,

$$g(v) = \frac{1}{r(\bar{i}(v))} \tag{9-10}$$

and

$$G(v) = \frac{1}{R(\bar{i}(v))} \tag{9-11}$$

For nonlinear capacitors and inductors there are definitions of capacitance, elastance, inductance, and reciprocal inductance that are analogous to the definitions of resistance and conductance. For a capacitor the variables are charge and voltage. For an inductor the quantities are flux linkage and current.

As a simple example, suppose we have a nonlinear resistor, inductor, and capacitor all in parallel as in Figure 9-3. If we choose v and i_L as state variables, we get

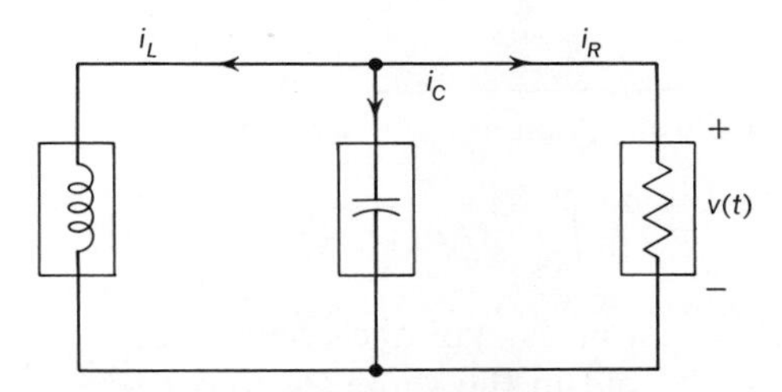

Figure 9-3 A simple nonlinear circuit.

$$\begin{aligned} &\frac{d}{dt}\bar{q}\,(v(t)) = -i_L - \bar{i}(v(t)) \\ &\frac{d}{dt}\bar{\lambda}\,(i_L(t)) = v(t) \end{aligned} \tag{9-12}$$

By the chain rule for differentiation

$$c(v)\,\dot{v} = i_L - \bar{i}(v)$$

$$l(i_L)\,\dot{i}_L = v$$

If the incremental capacitance and inductance are nonzero, these equations can be put in normal first-order form:

$$\begin{aligned} \dot{v} &= f_1(v, i_L) \\ \dot{i}_L &= f_2(v, i_L) \end{aligned} \tag{9-13}$$

where

$$f_1(v, i_L) = \frac{-i_L - \bar{i}(v)}{c(v)}$$

$$f_2(v, i_L) = \frac{v}{l(i_L)}$$

The integral equation formulation of nonlinear circuits is appropriate only when there are a few linear elements that can be combined into a box with a response function. To illustrate this formulation consider a nonlinear resistor in an otherwise linear, time-invariant circuit. By applying Thevenin's theorem to the linear, time-invariant part, as seen from the resistor terminals, we get the circuit of Figure 9-4.

By Kirchhoff's law:

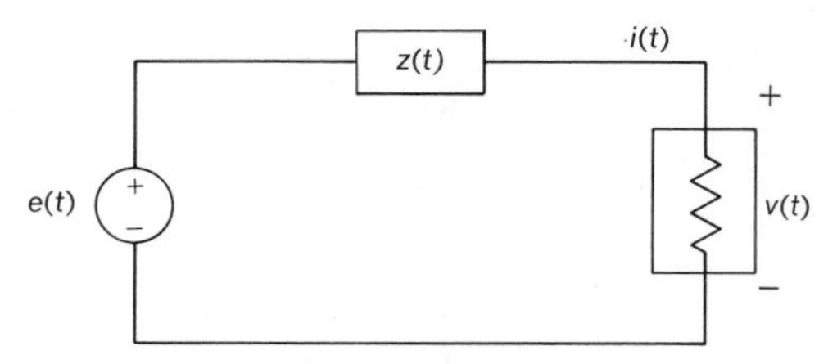

Figure 9-4 A circuit with only one nonlinearity.

$$e(t) = \int_{-\infty}^{t} z(t-\tau)\; i(\tau)\; d\tau + v(t) \tag{9-14}$$

If the nonlinear resistor is voltage-controlled, $i(t) = \bar{i}(v(t))$, then Equation (9-14) can be put in the same form as (9-2).

$$v(t) = e(t) - \int_{-\infty}^{t} z(t-\tau)\; \bar{i}(v(\tau))\; d\tau \tag{9-15}$$

As with the step from Equation (9-2) to (9-3), the lower limit on the integral in (9-15) can be taken as some finite time t_0 if the state of the linear network at t_0 is accounted for by appropriate terms in the Thevenin voltage source $e(t)$.

9-2 Second-Order Systems—Trajectories in the Phase Plane

When a nonlinear system is characterized by the n first-order equations (9-5), the vector $\mathbf{w}$ can be interpreted geometrically as a point in an n-dimensional space called *state space*. As a function of time $\mathbf{w}(t)$ traces a curve (or trajectory) in the state space. The basic geometric ideas are best illustrated by an example. The simplest meaningful example is a second-order system wherein the trajectory lies in a two-dimensional space and thus can be drawn on a sheet of paper.

As a second-order system example, consider the resonant tank shunted by a biased tunnel diode as shown in Figure 9-5. Since this

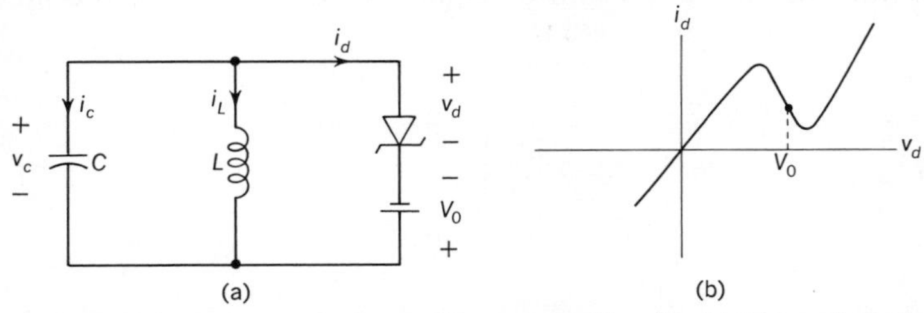

Figure 9-5 Tunnel diode oscillator. (**a**) Circuit. (**b**) Tunnel diode characteristics.

circuit is a special case of that of Figure 9-3, the normal form equations (9-13) apply. In this case they are

$$\dot{v}_C = -\frac{i_L + i_d}{C}$$
$$\dot{i}_L = \frac{v_C}{L} \tag{9-16}$$

The geometry of the solution of Equation (9-16) is most easily discussed by starting at an equilibrium point—a point where the equation is satisfied with both variables constant. If v_C is zero and $i_L = -\bar{i}_d(V_0)$, then both $\dot{v}_C$ and $\dot{i}_L$ are zero and Equation (9-16) is satisfied. This equilibrium point can be moved to the origin by the change of variables

$$x_1 = v_C$$
$$x_2 = i_L + \bar{i}_d(V_0) \tag{9-17}$$

The notation is further simplified if the nonlinear function $\bar{i}_d(v_d)$ is replaced by a new nonlinearity defined by

$$\bar{i}(x_1) = \frac{1}{C}\,[\bar{i}_d(x_1 + V_0) - \bar{i}_d(V_0)] \tag{9-18}$$

In terms of x_1, x_2, and $\bar{i}(x_1)$, the normal form equations (9-16) become

$$\dot{x}_1 = -\bar{i}(x_1) - \frac{1}{C}\,x_2$$
$$\dot{x}_2 = \frac{1}{L}x_1 \tag{9-19}$$

By the definition of $\bar{i}(x_1)$ in Equation (9-18) it is clear that $\bar{i}(0) = 0$. Thus $x_1 = 0, x_2 = 0$, is an equilibrium point of (9-19).

The behavior of the solution to Equations (9-19) near the origin will depend on the various circuit parameters including the bias voltage. To be specific, let us select the bias in the middle of the negative resistance region of the tunnel diode. Then $\bar{i}(x_1)$ is the heavy curve in Figure 9-6. Furthermore, let us take L and C equal to one to simplify the numbers.

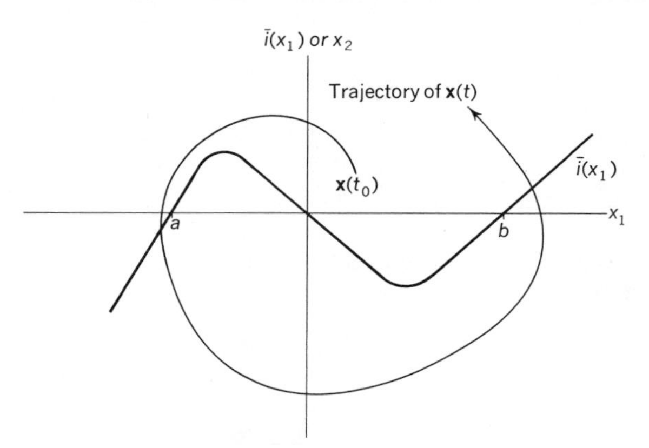

Figure 9-6 Phase plane trajectory.

Starting from some given initial condition, $x(t_0)$, the solution to Equations (9-19) can be constructed graphically. By approximating the derivatives by difference quotients, the value of $\mathbf{x}(t_0 + \Delta t)$ can be found from (9-19). Then $x(t_0 + 2\Delta t)$ can be found approximately, and so on. If the Δt increments are small enough, the curve connecting the points located by this procedure will be a very good approximation to the solution $x(t)$.

The light curve in Figure 9-6 is a trajectory plotted in the (x_1, x_2) plane. The point $\mathbf{x}(t_0)$ is such that x_1 and $\bar{i}(x_1)$ cancel one another in Equations (9-19). Thus, the trajectory starts with $\dot{x}_1 = 0, \dot{x}_2 = x_1$; that is, it is moving straight up. As x_2 increases, the trajectory moves toward the left. When it crosses the x_2 axis, x_1 and $\bar{i}(x_1)$ are both zero so the direction is horizontal. The curve is continued in the future until it has gone around one time. Then the length of the vector $\mathbf{x}$ is greater than the length of $\mathbf{x}(t_0)$. This trajectory is spiralling outward and the system appears to be unstable. If we continued the plot, we would see that the trajectory does not spiral out without bound, but that it approaches a limiting curve whose maximum distance from the origin is finite. Such a curve is called a *limit cycle*. These stability notions are discussed more precisely in the next section. The (x_1, x_2) plane for a second-order system is called the *phase plane*.

The length of the vector $\mathbf{x}$ is

$$\|\mathbf{x}\| = \sqrt{x_1{}^2 + x_2{}^2} \tag{9-20}$$

The derivative of the length is

$$\frac{d}{dt}\|\mathbf{x}\| = \frac{x_1\dot{x}_1 + x_2\dot{x}_2}{\|x\|} \tag{9-21}$$

Substituting for $\dot{x}_1$ and $\dot{x}_2$ from Equations (9-19) gives

$$\|\mathbf{x}\| = \frac{-x_1\bar{i}(x_1)}{\|x\|} \tag{9-22}$$

Since $\|\mathbf{x}\|$ is always positive, the length of $\mathbf{x}$ is increasing if x_1 and $\bar{i}(x_1)$ have opposite signs; it is decreasing when they have the same sign. Thus when the phase plane trajectory is superimposed on the $\bar{i}(x_1)$ characteristic as in Figure 9-6, the trajectory is spiralling outward when x_1 is between a and b and it is spiralling inward when $x_1 < b$ or $x_1 > a$.

The different types of equilibrium points, limit cycles, and so forth, are cataloged in many books on nonlinear systems or nonlinear oscillations (see References 15 and 25). These references also present many techniques for graphical construction of phase plane trajectories. Unfortunately most of these graphical methods apply only to second-order systems wherein the construction is made in a plane. For third- and higher-order systems the trajectory of the response is a curve in a three- or higher-dimensional space. The geometry of such curves is much more complex than the geometry of curves in a plane.

The two basic ideas presented above—the numerical integration of the differential equations and the concept of distance from the origin—extend to higher-order systems. Numerical integration is a useful method for solving specific system response problems for specific initial conditions. The application of distance from the origin to stability is the basic topic of the next section.

9-3 Some Additional Stability Notions

In nonlinear systems the notion of stability is more complicated than it is for linear systems. In linear, time-invariant systems the stability does not depend on the initial state except in the special case where one of the natural frequencies does not appear in the zero-input response for that initial state. A nonlinear system may be stable for some sets of initial conditions and unstable for others. In these cases it is not really meaningful to talk about the stability of the system. Instead, the idea of stability must be applied to classes of responses to initial conditions.

For a more precise definition of stability let us start with normal form differential equations that are a slight variation of Equation (9-5). That is,

$$\dot{\mathbf{x}} = \mathbf{F}(\mathbf{x}, t) \tag{9-23}$$

In Equation (9-23) the nonlinear function $\mathbf{F}$ is an n-vector-valued function of the $n + 1$ variables $x_1, x_2, \cdots, x_n, t$. Thus, Equation (9-23) includes equations with variable coefficients with or without a forcing term as well as those where the right side does not depend on t explicitly.[2] Since (9-23) contains a description of both the system and its inputs, the state at t_0 plus the given equation determines the solution $x(t)$ for $t > t_0$. This is the definition of the state at t_0. For the present discussion we assume that the state at t_0 is the vector $\mathbf{x}(t_0)$. In Chapter 10 we shall see that a very large class of nonlinear systems satisfies this assumption. For notational purposes we let $\mathbf{x}(t, \mathbf{x}(t_0))$ designate the values of $\mathbf{x}(t)$ for given $\mathbf{x}(t_0), t \geqslant t_0$.

For the system (9-23), a response $\mathbf{x}(t, \mathbf{x}(t_0))$ is said to be *stable in the Lyapunov sense* if for every $\epsilon > 0$ there exists a $\delta > 0$ such that

$$\|\mathbf{x}(t, \mathbf{x}(t_0)) - \mathbf{x}(t, \mathbf{x}_1(t_0))\| < \epsilon \text{ for all } t > t_0$$

whenever

$$\|\mathbf{x}(t_0) - \mathbf{x}_1(t_0)\| < \delta \tag{9-24}$$

where the double bars enclosing a vector means the length as defined for a two-dimensional vector by Equation (9-20).[3]

Since we cannot, in general, get explicit formulas for the response of nonlinear systems characterized by equations such as (9-23), it is extremely difficult to check the definition of Lyapunov stability directly. On the other hand, we do have formulas for the time rate of change of the response. These formulas are the normal form differential equations. Thus, although we cannot compute the response explicitly, we can see if it is getting smaller or larger.

By the use of the n-dimensional generalization of Equation (9-21), we can easily find some sufficient conditions that guarantee the stability of the origin in state space. From Equations (9-21) and (9-23), this generalization is

$$\frac{d}{dt}\|\mathbf{x}\| = \frac{\mathbf{x}^T\mathbf{F}(\mathbf{x}, t)}{\|\mathbf{x}\|} \tag{9-25}$$

The numerator on the right of Equation (9-25) is a function of the $(n + 1)$ variables. If the values of this function are negative for all $\mathbf{x}$ in an n-dimensional sphere of radius N in state space and for all $t > t_0$, then starting from any initial state $\mathbf{x}(t_0)$ for which $\|\mathbf{x}(t_0)\| < N$, $\mathbf{x}(t)$ will approach the

[2] For studying certain properties of Equation (9-22) that are beyond the scope of this text, the explicit dependence of the right side on t is an important property. When t does not appear, the system is said to be *autonomous*. Otherwise it is *nonautonomous*.

[3] The generalization of length to any norm can be used in the definition, but a discussion of norms is not necessary for the development below.

origin as $t \to \infty$. In terms of the symbols of the stability definition (9-24), the correspondence of the situation just described is as follows:

$$\mathbf{x}(t, \mathbf{x}(t_0)) = \mathbf{0} - \text{the zero vector}$$

$$\delta = \epsilon < N$$

For this special case we have the following:

If the function $\mathbf{x}^T\mathbf{F}$ is negative definite for all $\|\mathbf{x}\| < N, t > t_0$, the origin in state space is stable in the sense of Lyapunov. Furthermore, if N can be infinite, the system is asymptotically stable as defined in Chapter 8.

As an example consider the system of Figure 9-7. The system con-

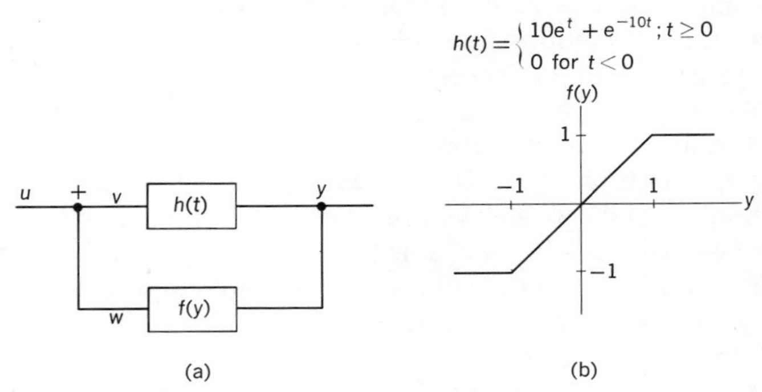

Figure 9-7 A stability problem.

sists of an unstable forward loop stabilized by negative feedback through a saturating amplifier. The equations of the system are

$$\begin{aligned} \dot{x}_1 &= x_1 + 10v \\ \dot{x}_2 &= -10x_2 + v \\ y &= x_1 + x_2 \\ v &= u - f(y) \end{aligned} \qquad \textbf{(9-26)}$$

Combining these equations to eliminate v gives

$$\begin{aligned} \dot{x}_1 &= x_1 - 10f(x_1 + x_2) + 10u \\ \dot{x}_2 &= -10x_2 - f(x_1 + x_2) + u \\ y &= x_1 + x_2 \end{aligned} \qquad \textbf{(9-27)}$$

When $u(t) = 0$, $\mathbf{x}$, the zero vector, is a solution. For any other solution, $\mathbf{x}_1(t)$, starting from some $\mathbf{x}_1(t_0)$, we can see whether this solution approaches the zero vector by examining the numerator on the right side of Equation (9-25). In this case, that numerator is

$$\mathbf{x}^T\mathbf{F} = x_1^2 - 10x_1 f(x_1 + x_2) - 10x_2^2 - x_2 f(x_1 + x_2) \qquad \textbf{(9-28)}$$

For $|x_1 + x_2| < 1$ we have

$$\mathbf{x}^T\mathbf{F} = -9x_1^2 - 11x_1x_2 - 11x_2^2 \tag{9-29}$$

Since the matrix $\begin{bmatrix} 9 & 11/2 \\ 11/2 & 11 \end{bmatrix}$ forms a positive definite quadratic form, $\mathbf{x}^T\mathbf{F}$ is negative definite. Thus, for $|x_1 + x_2| < 1$, the zero-vector solution is stable in the sense of Lyapunov if δ in the definition (9-24) is taken as ϵ when $\epsilon < 0.707$ and as 0.707 when $\epsilon > 0.707$. For any initial condition with both $x_1(t_0)$ and $x_2(t_0)$ positive and $(x_1(t_0) + x_2(t_0)) > 1$ the system is unstable. Then the effect of the feedback is merely to put a constant signal of amplitude one at the input of the unstable open-loop system. For other initial conditions, the solution may get into the stable region. Then the response approaches zero as $t \to \infty$.

The above ideas on the stability of the origin were generalized by Lyapunov in an important theorem known as the *second method*, or the *direct method, of Lyapunov*. In the development above we started with the length of $\mathbf{x}$, which is always positive, and then examined its derivative. If the derivative was always negative we concluded the system was stable. From a physical argument we could have equally well examined a system with positive energy storage and considered the rate of change of stored energy. In an undriven system the discussion of the previous chapter showed that this rate of change is minus the dissipation function, P_d. Thus if P_d is positive definite the system is stable. Lyapunov showed that there was a whole class of functions of the state variables and time that could be considered in addition to the length of $\mathbf{x}$ and the stored energy. He called these functions *V-functions*, defined as follows:

For a system described by Equation (9-23), a V-function, $V(\mathbf{x}, t)$, is any scalar function that is real and continuous in a sphere of radius N of state space and for $t > t_0$. Furthermore, its first partial derivatives with respect to the x_i and t must be continuous and $V(\mathbf{0}, t) = 0$ for all $t > t_0$.

The basic theorem (there are several variations) states: If a V-function exists which is definite (positive or negative) and $\dot{V}$ is a definite function of the opposite sign, then starting from any initial state $\mathbf{x}(t_1)$, $t_1 > t_0$ and $\|\mathbf{x}(t_1)\| < N$, $\mathbf{x}(t) \to \mathbf{0}$ as $t \to \infty$.

The proof of the theorem in general is beyond the scope of this text (see Reference 4). For the problems at the end of this chapter either the dissipative energy or $\|\mathbf{x}\|$ will be adequate as a V-function.

Stability of other solutions. The above discussion has concerned only the stability of the origin as a solution to equations of unforced systems. For the more general cases of Equation (9-23), the above ideas can also be used to investigate the stability of other solutions. The discussion will be made more complete after we obtain some results from the integral equation formulation of the next chapter. To begin, consider the solution

$\mathbf{x}(t, \mathbf{x}(t_0))$ to Equation (9-23). We should like to examine the relation between $\mathbf{x}(t, \mathbf{x}(t_0))$ and a second solution $\mathbf{x}_1(t, \mathbf{x}_1(t_0))$ that has a different state at t_0. Specifically we should like to know if the trajectories of these two solutions stay close together in state space if they start close together. Since the basic techniques apply to a first-order nonlinear equation as well as higher-order systems, we begin with the one-dimensional case:

Consider the equation

$$\dot{x} = f(x, t) \tag{9-30}$$

Suppose $x_0(t)$ and $x_1(t)$ are both solutions to this equation. Then taking the difference between Equation (9-30) with $x_0(t)$ and (9-30) with $x_1(t)$ gives

$$\dot{x}_1(t) - \dot{x}_0(t) = f(x_1, t) - f(x_0, t) \tag{9-31}$$

If $f(x, t)$ has a continuous second partial derivative with respect to x, we can apply the differential approximation theorem.[4] In this case the theorem yields

$$f(x_1, t) = f(x_0, t) + \frac{\partial f(x_0, t)}{\partial x} [x_1(t) - x_0(t)] + R(x_1 - x_0) \tag{9-32}$$

where the remainder, $R(x_1 - x_0)$, goes to zero faster than $(x_1 - x_0)$. That is,

$$\underbrace{\lim}_{(x_1 - x_0) \to 0} \quad \frac{R(x_1 - x_0)}{x_1 - x_0} = 0 \tag{9-33}$$

Using Equation (9-32) in (9-31) and using Δ for $(x_1 - x_0)$ gives

$$\dot{\Delta} = \frac{\partial f(x_0, t)}{\partial x} \Delta + R(\Delta) \tag{9-34}$$

This equation is a nonlinear differential equation in Δ. Since we are interested in solutions x_1 that start and remain near x_0, we are interested in Equation (9-34) near the origin. Furthermore, in cases where we can show that the effect of $R(\Delta)$ is negligible, Equation (9-3) becomes a linear equation with a variable coefficient. The coefficient $\partial f(x_0, t)/\partial t$ is a known time function; since presumably $x_0(t)$ is known for all $t > t_0$, and we are investigating its stability.

The generalization to higher-order equations is no more than the generalization of Taylor's theorem to functions of several variables. For a vector-valued function $\mathbf{F}(\mathbf{x})$ there are n^2 partial derivatives. For each of the n functions, there are partials for each variable. The array of partials is called the *Jacobian matrix*. We denote this matrix by $\partial \mathbf{F}/\partial \mathbf{x}$. Specifically, $\mathbf{x}$ and $\mathbf{F}$ are n-vectors that can be written more explicitly as

[4] The differential approximation theorem is just Taylor's theorem with only one differential term.

$$\mathbf{x}(t) = \begin{bmatrix} x_1(t) \\ x_2(t) \\ \vdots \\ x_n(t) \end{bmatrix} \tag{9-35}$$

$$\mathbf{F}(\mathbf{x}, t) = \begin{bmatrix} f_1(x_1, x_2, \cdots, x_n, t) \\ f_2(x_1, x_2, \cdots, x_n, t) \\ \vdots \\ f_n(x_1, x_2, \cdots, x_n, t) \end{bmatrix} \tag{9-36}$$

Then the Jacobian matrix is

$$\frac{\partial \mathbf{F}}{\partial \mathbf{x}} = \begin{bmatrix} \frac{\partial f_1}{\partial x_1} & \frac{\partial f_2}{\partial x_1} & \cdots & \frac{\partial f_n}{\partial x_1} \\ \frac{\partial f_1}{\partial x_2} & \frac{\partial f_2}{\partial x_2} & \cdots & \frac{\partial f_n}{\partial x_2} \\ \vdots & \vdots & & \vdots \\ \frac{\partial f_1}{\partial x_n} & \frac{\partial f_2}{\partial x_n} & & \frac{\partial f_n}{\partial x_n} \end{bmatrix} \tag{9-37}$$

The procedure for deriving an n-dimensional differential equation that gives the differences between the components of a given solution $\mathbf{x}_0(t)$ and a second solution $\mathbf{x}_1(t)$ is the same for the first-order case. Each of the n equations involves a row of the Jacobian matrix (9-37). There is a remainder with each equation. These can be represented by a remainder vector $\mathbf{R}$. If we now use Δ to represent an n-vector of differences between $\mathbf{x}_0$ and $\mathbf{x}_1$, the generalization of (9-34) to a study of the stability of a solution $\mathbf{x}_0$ to (9-23) is

$$\dot{\Delta} = \frac{\partial \mathbf{F}}{\partial \mathbf{x}} \Delta + \mathbf{R}(\Delta) \tag{9-38}$$

We shall return to (9-38) after obtaining some results that are most easily derived via integral equations. These results deal with estimates of both $\mathbf{R}(\Delta)$ and Δ. They are the primary objectives of the next chapter.

PROBLEMS

9-1 For the block diagram of Figure P9-1 let $h(t)$ represent a causal system with response for $t > 0$ as $e^{-t} - e^{-2t}$. Let the nonlinear function be

$$f(x) = \tfrac{1}{2} x \left(1 + \tfrac{1}{4} x^2\right)$$

Let the input $u(t)$ be a pulse of height one, duration 2 s, starting at $t = 0$. The state at $t = 0$ is zero.

a. Set up the problem as an integral equation in the form (9-3).
b. Set up normal form equations in the form (9-5). Be sure to define **w** and **F** in terms of the given quantities.

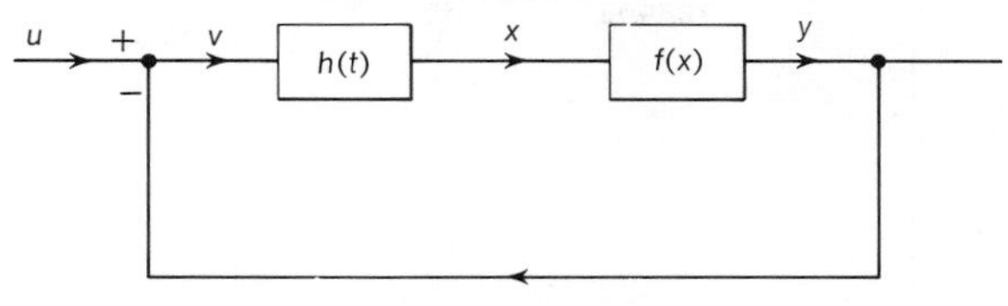

Figure P9-1

9-2 Consider the block diagram of Figure P9-2. Let the linear system be characterized by the $\underset{\sim}{A}$ matrix $\begin{bmatrix} -4 & 1 \\ -6 & 1 \end{bmatrix}$, **B** vector $\begin{bmatrix} 1 \\ 0 \end{bmatrix}$, $\underset{\sim}{C}$ matrix $[0 \quad 1]$, and $\underset{\sim}{D}$ and $\underset{\sim}{E}$ matrixes zero. The nonlinear function is

$$f(y) = \tfrac{1}{2}\, y\, (1 + \tfrac{1}{4}\, y^2)$$

Let $\mathbf{x}(0) = \begin{bmatrix} 1 \\ 1 \end{bmatrix}$ for the linear system denoted by $h(t)$ in the diagram. The input is zero for $t > 0$.
a. Set up the problem as an integral equation in the form (9-3).
b. Set up normal form equations in the form (9-5).

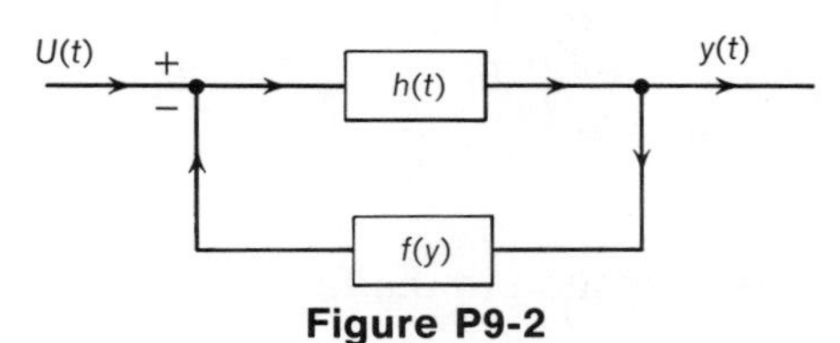

Figure P9-2

9-3 Repeat Problem 9-2 for the block diagram of Figure P9-3. Use the same $\underset{\sim}{A}$, $\underset{\sim}{B}$, and $\underset{\sim}{C}$ matrixes and the same functional form for the nonlinearity.

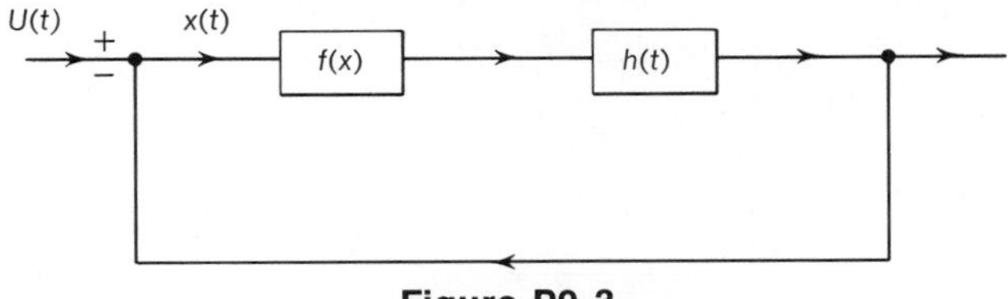

Figure P9-3

9-4 Consider the circuit of Figure P9-4 with the nonlinear resistance characterized by the voltage-current relationship

$$\overline{v}(i) = i^3 - i$$

For a general input voltage $e(t)$:

a. Set up the problem as an integral equation in the form (9-3).
b. Set up normal form equations in the form (9-5).

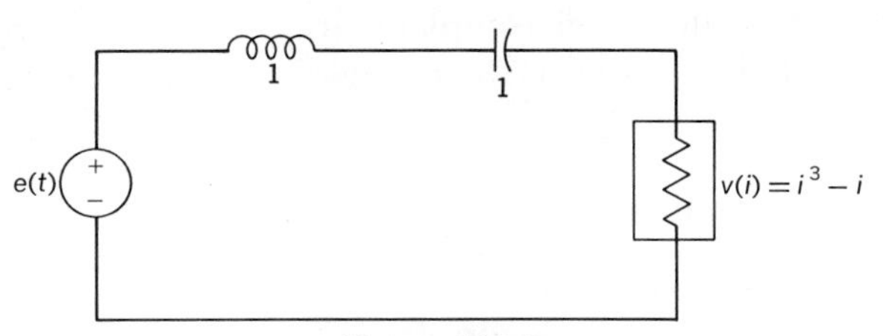

Figure P9-4

9-5 Consider the circuit of Figure P9-5 with the time-variant capacitor

$$C(t) = 4 + 2\cos t$$

Let the admittance matrix of the box be

$$\begin{bmatrix} 2 + \dfrac{1}{j\omega} & -\dfrac{1}{j\omega} \\ -\dfrac{1}{j\omega} & 1 + \dfrac{1}{j\omega} \end{bmatrix}$$

Let

$$e(t) = \begin{cases} 0 & \text{for } t < 0 \\ \cos \dfrac{t}{2} & \text{for } t \geqslant 0 \end{cases}$$

a. Set up normal form equation in the form

$$\mathbf{x} = \underset{\sim}{A}(t)\,\mathbf{x} + \mathbf{B}\,e(t)$$

$$v_2(t) = \underset{\sim}{C}\,\mathbf{x}$$

where the matrix $\underset{\sim}{A}(t)$ is a matrix of functions of time.
b. Set up an integral equation in the form

$$v_2(t) = g(t) + \int_0^t h(t, \tau)\, v_2(\tau)\, d\tau$$

where $g(t)$ and $h(t, \tau)$ are determined from the given information.

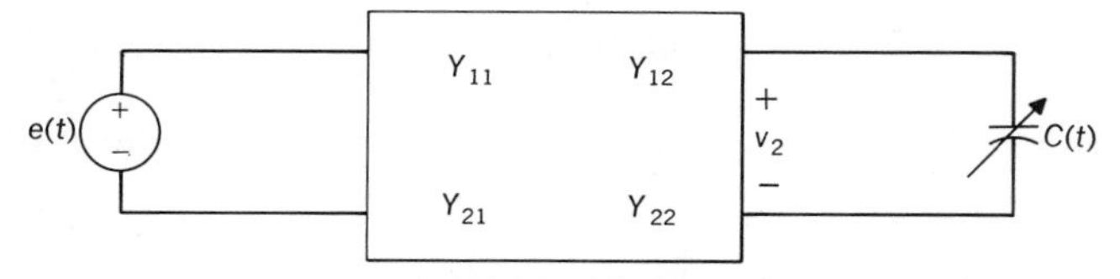

Figure P9-5

9-6 Consider the circuit of Problem 9-4.

a. Find the limit cycle.

b. Carefully construct a phase plane trajectory going once around starting from a point in the unstable region and another starting from a point in the stable region.

9-7 Consider the feedback diagram of Figure 9-1. Let

$$h(t) = e^{-t} - e^{-2t}$$

$$f(x) = \begin{cases} K_1 x & \text{for } |x| < 1 \\ K_2 x + K_1 - K_2 & \text{for } |x| > 1 \end{cases}$$

a. Select K_1 and K_2 so that the system has a limit cycle.

b. Show that if

$$h(t) = e^{-t} + e^{-2t}$$

and $f(x)$ is as above, there is no limit cycle when K_1 and K_2 are positive.

9-8 Consider the circuit of Figure P9-8. Show that a sufficient condition for asymptotic stability of the origin for all initial conditions is that the $v - i$ characteristic of the nonlinear resistor lies in the first and third quadrant.

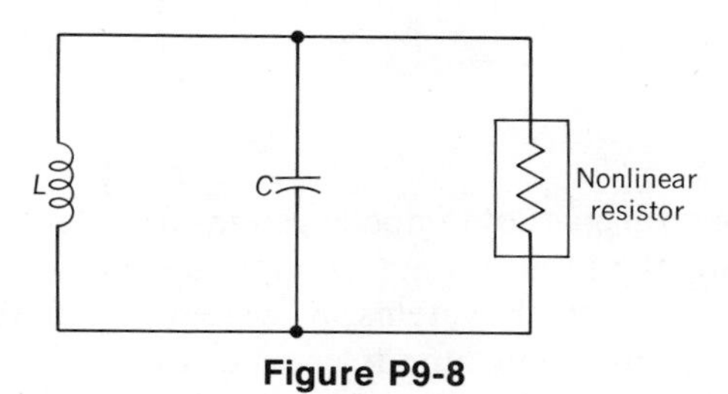

Figure P9-8

9-9 Consider the feedback diagram of Figure 9-1. Suppose the linear system is described by the matrix differential equations plus input output statement of Equation (6-1). Find a set of sufficient conditions on the various matrixes $\underset{\sim}{A}$, $\underset{\sim}{B}$, $\underset{\sim}{C}$, $\underset{\sim}{D}$, $\underset{\sim}{E}$, and the nonlinear function $f(x)$ that guarantee the asymptotic stability of the origin for all initial conditions when the input is zero.

CHAPTER

10

Iteration of Nonlinear Systems Problems

The one fairly general class of lumped systems for which explicit formulas can be derived for the input-output relation are the linear, time-invariant systems. For other lumped systems, a system of differential or integral equations relating input and output can be formulated; but there is no way to construct the solution to the equation as an explicit formula. There are, of course, methods for constructing a specific solution to a specific equation. One of these methods is numerical integration. Another is the method of iteration, a successive approximation technique. This latter method results in a sequence of formulas that approximate the response. These formulas can be used to prove certain important system properties and to get bounds on the response of a system to specific inputs. The method of iteration can also be executed numerically on a digital computer to give a specific solution to a specific system analysis problem. In this chapter we use the method of iteration to derive some properties of systems and to obtain some bounds on system outputs as a function of bounds on the inputs. No serious consideration of numerical methods is included.

10-1 The State at t_0

In Chapter 3 it was shown that for linear systems of order n the state at t_0 was an n-vector of numbers. In this section, we show that for a nonlinear system in the form of Figure 9-1 of Chapter 9 the state at t_0 is just the state of the linear, time-invariant part when the nonlinear function f is continuously differentiable.[1]

For the feedback system of Figure 9-1, Equation (9-3) is the corresponding integral equation. To show how the state at t_0 can be included in the equation, let us reformulate Equation (9-3) without the assumption that $u(t)$ is zero for $t < t_0$. The quantity x in Figure 9-1 is the output of a linear, time-invariant system. At any time $t > t_0$,

$$x(t) = \int_{t_0}^{t} h(t-\tau)\ v(\tau) d\tau + g_1(t) \qquad \textbf{(10-1)}$$

where $g_1(t)$ is the zero-input response due to the state at t_0 of the linear system characterized by $h(t)$. If this linear system is nth order, the state is given by an n-vector of numbers.

From the algebraic rules of the feedback diagram of Figure 9-1, $v(t)$ in Equation (10-1) can be written in terms of $u(t)$ and $f(x(t))$. Thus the equation for the system is

$$x(t) = \int_{t_0}^{t} h(t-\tau)\ u(\tau)\ d\tau + g_1(t) - \int_{t_0}^{t} h(t-\tau)\ f(x(\tau))\ d\tau \qquad \textbf{(10-2)}$$

This equation is still in the form of (9-3). For any $t > t_0$, it has a unique solution when the nonlinear function f is continuously differentiable. The complete proof is in Reference 27. Some of the more important steps of the proof are discussed below.

Construction of a sequence of iterates. The functional form of Equation (10-2) is still that of (9-3). Thus our basic equation is

$$x(t) = g(t) - \int_{t_0}^{t} h(t-\tau)\ f(x(\tau))\ d\tau \qquad \textbf{(10-3)}$$

The solution to this equation can be constructed by successive approximations. The particular technique for generating these approximations is called *iteration*. The functions defined at the various steps of the procedure are called *iterates*. The first iterate $x_1(t)$ is obtained by assuming that the nonlinear element is absent and that $x_1(t) = g(t)$. The second iterate is then obtained by passing the first iterate through the f

[1] The result is valid for some functions that are not continuously differentiable. To introduce the added generality requires a bit more mathematical background than assumed for this text. Reference 27 and its bibliography discuss the mathematical problem in general.

box in Figure 9-1, feeding this result back and through h, and adding to the results of the input and initial conditions in the response of h. The third iterate is then obtained in the same manner by feeding the second iterate around through f and h, and so on.

Formally, starting from Equation (10-3)

$$\begin{aligned} x_1(t) &= \int_{t_0}^{t} h(t-\tau)\, u(\tau)\, d\tau + g_1(t) = g(t) \\ x_2(t) &= g(t) - \int_{t_0}^{t} h(t-\tau)\, f(x_1(\tau))\, d\tau \\ x_3(t) &= g(t) - \int_{t_0}^{t} h(t-\tau)\, f(x_2(\tau))\, d\tau \\ &\vdots \\ x_n(t) &= g(t) - \int_{t_0}^{t} h(t-\tau)\, f(x_{n-1}(\tau))\, d\tau \\ &\vdots \end{aligned} \tag{10-4}$$

Each of the successive $x_n(t)$'s is computable in terms of known quantities after the previous iterates are computed. Thus Equation (10-4) defines an infinite sequence of functions. We must show that this sequence converges; that the function defined by the limit as $n \to \infty$ of the sequence is $x(t)$, the solution to Equation (10-3); and that this is the only solution to Equation (10-3) for all $t > t_0$. Since the object of this text is the presentation of basic techniques with a minimum emphasis on special cases, the class of nonlinear functions f and linear system operators h is restricted to those that can be considered with the methods of elementary calculus. In Reference 27, a more general class of functions and operators is considered and shown to be within the capabilities of the methods described below.

For the present, let us restrict the nonlinear function to one which has a continuous, uniformly bounded derivative. That is, for all x, $f'(x)$ is defined, continuous, and bounded by $\pm F$, F being a positive number. The linear system whose response function is h is lumped, causal, asymptotically stable, and time-invariant. Thus, $h(t)$ is a sum of decaying exponentials, or polynomials times such exponentials. With these conditions on f and h, certain operations needed to prove that Equation (10-4) generates a unique solution to (10-3) are easily shown.

Since f is continuously differentiable, the mean-value theorem applies. Specifically,

$$f(x) - f(y) = f'(z)\,[x - y] \tag{10-5}$$

where $x \leqslant z \leqslant y$.

Furthermore, since $f'(x)$ is uniformly bounded by $\pm F$, the following estimate[2] applies

$$|f(x) - f(y)| \leqslant F\,|x - y| \tag{10-6}$$

With the linear system restricted as described above, $h(t)$ is absolutely integrable and square integrable on the interval $(0, \infty)$.

That is,

$$\int_0^\infty |h(t)|\,dt = H_1 < \infty \tag{10-7}$$

and

$$\int_0^\infty h^2(t)\,dt = H_2 < \infty \tag{10-8}$$

where H_1 and H_2, like F in Equation (10-6), are positive numbers.

The most convenient way to show convergence of a sequence of functions such as (10-4) is to use the well-known methods for showing convergence of infinite series. The first step is to find a series whose successive partial sums are the successive members of the sequence under investigation. One possibility is to define a new set of functions by successive differences of the x_n. That is,

$$\begin{aligned} D_1(t) &= x_1(t) \\ D_2(t) &= x_2(t) - x_1(t) \\ &\vdots \\ D_n(t) &= x_n(t) - x_{n-1}(t) \\ &\vdots \end{aligned} \tag{10-9}$$

Then through successive cancellations in any sum of $D_n(t)$,

$$x_n(t) = \sum_{k=1}^{n} D_k(t) \tag{10-10}$$

Now for all $t \geqslant t_0$ the uniform convergence of the infinite series of $D_k(t)$ is equivalent to the uniform convergence of the sequence of $x_n(t)$. Thus the first problem is to show the uniform convergence of $\Sigma_{k=1}^{\infty} D_k(t)$.

Demonstration of convergence of the iterates. The procedure for showing the uniform convergence of a series of functions is to first find

[2] Formula (10-6) is known as a Lipschitz condition on the function f. As long as there is a number F for which (10-6) is satisfied the sequence (10-4) converges to the desired solution. The continuity of f' is not necessary. It is certainly sufficient.

a series of numbers such that each number is a uniform upper bound on the corresponding function. Then the well-known tests (see, for example, Reference 24, Chapter 2) for convergence of infinite series of numbers can be applied. The estimates (10-6), (10-7), and (10-8) can be used to get upper bounds on the $D_k(t)$. In addition, one more well-known theorem, Schwartz's inequality, is needed. For integrals such as those in Equation (10-3), this inequality becomes

$$\left[\int_a^b y(t)\, z(t) dt\right]^2 \leqslant \left[\int_a^b y^2(t)\, dt\right]\left[\int_a^b z^2(t)\, dt\right] \tag{10-11}$$

Substituting Equation (10-4) into (10-9) and regrouping gives the general form for $D_n(t)$ as

$$D_n(t) = \int_{t_0}^t h(t-\tau)\, [f(x_{n-2}(\tau)) - f(x_{n-1}(\tau))]\, d\tau \tag{10-12}$$

Applying Inequality (10-11) to estimate the right side of Equation (10-12) gives

$$D_n^2(t) \leqslant \left[\int_{t_0}^t h^2(t-\tau)\, d\tau\right]\left[\int_{t_0}^t [f(x_{n-2}(\tau)) - f(x_{n-1}(\tau))]^2\, d\tau\right] \tag{10-13}$$

For the first bracket on the right of (10-13), a change of variables gives

$$\int_{t_0}^t h^2(t-\tau)\, d\tau = \int_0^{t-t_0} h^2(\tau)\, d\tau \tag{10-14}$$

Since $h^2(t)$ is positive, the integral is a monotonic function of the upper limit. Thus with the symbols of Equation (10-8)

$$\int_0^{t-t_0} h^2(\tau)\, d\tau \leqslant \int_0^\infty h^2(\tau)\, d\tau = H_2 \tag{10-15}$$

For the second integral in (10-13), the mean-value theorem can be used to estimate the integrand. From Inequality (10-6)

$$[f(x_{n-2}(\tau)) - f(x_{n-1}(\tau))]^2 \leqslant F^2[x_{n-2}(\tau) - x_{n-1}(\tau)]^2 \tag{10-16}$$

But the bracket on the right of Inequality (10-16) is just $D_{n-1}^2(\tau)$. Consequently, (10-13) along with (10-15) and (10-16) give the estimate

$$D_n^2(t) \leqslant H_2 F^2 \int_{t_0}^t D_{n-1}^2(\tau)\, d\tau \tag{10-17}$$

A general formula for estimating $D_n^2(t)$ in terms of known quantities can be obtained from Inequality (10-17) by starting with $n = 2$ and working up. Thus

$$D_2^2(t) \leqslant H_2\, F^2 \int_{t_0}^t g^2(\tau)\, d\tau \tag{10-18}$$

Since $g(t)$ is the total response of an asymptotically stable lumped, linear system, it is uniformly bounded provided the input is bounded.

$$G = \max g(\tau); \quad \tau > t_0 \tag{10-19}$$

Taking Inequality (10-18) and Equation (10-19) together gives

$$D_2^2(t) \leqslant H_2 F^2 G^2 (t - t_0) \tag{10-20}$$

Substituting Inequality (10-20) into (10-17) for $n = 3$ gives

$$D_3^2(t) \leqslant H_2^2 F^4 G^2 \int_{t_0}^{t} (\tau - t_0)\, d\tau = H_2^2 F^4 G (t - t_0)^2 \tag{10-21}$$

Substituting Inequality (10-21) into (10-17) for $n = 4$ gives

$$D_4^2(t) \leqslant H_2^3 F^6 G^2 \int_{t_0}^{t} (\tau - t_0)^2\, d\tau = H_2^3 F^6 G^2 \frac{(t - t_0)^3}{2} \tag{10-22}$$

From (10-22) the general formula is readily seen to be

$$D_n^2(t) \leqslant H_2^{(n-1)} F^{2(n-1)} G^2 \frac{(t - t_0)^{n-1}}{(n-2)!} \tag{10-23}$$

Since the right side of Inequality (10-23) is positive, there is no problem taking square roots. Thus,

$$|D_n(t)| \leqslant \frac{G}{FH_2 \sqrt{(t - t_0)}} \quad \frac{F^n (\sqrt{H_2})^n (t - t_0)^{n/2}}{\sqrt{(n-2)!}} \tag{10-24}$$

Since the right side of Inequality (10-24) is a monotone increasing function of t, a constant upper bound for all $t < T_0$ is

$$|D_n(t)| \leqslant A \frac{B^n}{\sqrt{(n-2)!}} \tag{10-25}$$

where

$$A = \frac{G}{F\sqrt{H_2(T_0 - t_0)}}$$

$$B = F\sqrt{H_2(T_0 - t_0)}$$

The problem of proving the uniform convergence of the sequence (10-3) has been reduced to proving the convergence of the series

$$\sum_{n=1}^{\infty} \frac{B^n}{\sqrt{(n-2)!}}$$

For the ratio test we examine

$$\lim_{n \to \infty} \frac{B^n/\sqrt{(n-2)!}}{B^{n-1}/\sqrt{(n-3)!}} = \lim_{n \to \infty} B/\sqrt{n-2} = 0 \tag{10-26}$$

Thus the series of numbers converges absolutely, and the sequence of functions $x_n(t)$ converges uniformly and absolutely for all $t > t_0$.

Uniqueness of the solution. The proof that the function $x(t)$ defined by the limit of the sequence (10-4) is also a solution to the original Equation (10-3) requires a bit more mathematical subtlety than the assumed background for this text. The essence of the proof is that although x_n and x_{n-1} appear, respectively, on the left and right sides of the successive equations in (10-4), the error introduced by using the limit function x on both sides becomes smaller and smaller as n takes on larger and larger values. In fact, it can be shown that this error goes to zero as n approaches infinity and the result is the original Equation (10-3).

To show that (10-3) has a unique solution, we assume that there are two functions $\bar{x}(t)$ and $x(t)$ that satisfy the equation and then show that the difference $\bar{x}(t) - x(t)$ must be zero for all t. The function $x(t)$ defined by the sequence (10-4) is bounded so long as $g(t)$ is bounded on the interval of interest. There may be a second solution $\bar{x}(t)$ to Equation (10-3) that is not well-behaved. For the proof below we show only that there is no solution that is not at least-square integrable on the interval (t_0, T_0).

The starting point for the proof is the assumption that both $x(t)$ and $\bar{x}(t)$ satisfy Equation (10-3). Then we can write

$$x(t) = g(t) - \int_{t_0}^{t} h(t-\tau)\, f(x(\tau))\, d\tau \tag{10-27a}$$

and

$$\bar{x}(t) = g(t) - \int_{t_0}^{t} h(t-\tau)\, f(\bar{x}(\tau))\, d\tau \tag{10-27b}$$

Subtracting Equation (10-27b) from (10-27a) gives

$$x(t) - \bar{x}(t) = \int_{t_0}^{t} h(t-\tau)\, [f(\bar{x}(\tau)) - f(x(\tau))]\, d\tau \tag{10-28}$$

Since Equation (10-28) is similar to (10-12), the steps that led to (10-18) give

$$(x(t) - \bar{x}(t))^2 \leqslant H_2 F^2 \int_{t_0}^{t} (x(\tau) - \bar{x}(\tau))^2\, d\tau \tag{10-29}$$

Since by assumption $x(t)$ and $\bar{x}(t)$ are both square integrable on (t_0, T_0), for $t < T_0$

$$\int_{t_0}^{t} (x(\tau) - \bar{x}(\tau))^2\, d\tau \leqslant \int_{t_0}^{T_0} (x(\tau) - \bar{x}(\tau))^2\, d\tau \leqslant K \tag{10-30}$$

with K a positive number. Combining Inequality (10-30) with (10-29) gives

$$(x(t) - \bar{x}(t))^2 \leqslant H_2 F^2 K \tag{10-31}$$

Now if Inequality (10-31) is an upper bound on the integrand on the right of (10-29), the result is

$$(x(t) - \bar{x}(t))^2 \leqslant H_2^2 F^4 K \int_{t_0}^{t} d\tau \tag{10-32}$$

Repeating the procedure with Inequality (10-32) instead of (10-31) gives

$$(x(t) - \bar{x}(t))^2 \leqslant H_2^3 F^6 K \int_{t_0}^{t} \left[\int_{t_0}^{\tau} d\lambda \right] d\tau \leqslant H_2^3 F^6 K \frac{(T_0 - t_0)^2}{2} \tag{10-33}$$

Successive repetition of this step leads to the general form

$$(x(t) - \bar{x}(t))^2 \leqslant K \frac{(H_2 F^2 (T_0 - t_0))^n}{n!} \tag{10-34}$$

Taking the limit of Inequality (10-34) as $n \to \infty$ shows that $(x(t) - \bar{x}(t))^2$ is zero. Thus $x(t)$ is the unique well-behaved solution for all $t < -T_0, -T_0$ an arbitrary number.

10-2 Bounds on the Response

For an engineer, the discussion of the previous section shows that the feedback system of Figure 9-1, or the nonlinear circuit of Figure 9-4, gives a unique bounded output at any time t, if the state of the linear part of the system is well-defined at some t_0 and the input is bounded from t_0 to t.[3] Furthermore, the iteration scheme (10-4) gives an algorithm for computing the response to as good an approximation as desired within the limitations of computer precision and time required for performing the successive steps. In many control applications, this constitutes a solution to the engineering problem.

Failure of the proof of Section 10-1 for steady state. On the other hand, communications and circuits problems are often essentially steady-state problems. For a steady-state, the most convenient engineering solution is one for finite t, but with the initial state taken at $-\infty$. The demonstration of convergence of the iterates above does not apply as $t_0 \to \infty$. The bound (10-24) requires $(t - t_0)$ to be finite. Since the bound is valid for any finite t_0, we might consider letting $n \to \infty$ first, and then

[3] In the discussion these attributes were found for x rather than the response y. However, if x in Figure 9-1 is known, $y = f(x)$ is easily obtained and it has the same properties.

letting $t_0 \to -\infty$. This will not work either as the following argument demonstrates. The terms of the bounding series given by the right side of Inequality (10-24) have essentially the form of the square root of the terms of the power series expansion of the exponential. When $A_k \geqslant 0$

$$\sqrt{\sum_{k=1}^{n} A_k} \leqslant \sum_{k=1}^{n} \sqrt{A_k} \tag{10-35}$$

Thus as $t_0 \to -\infty$ the bounding series grows faster than $e^{F^2 H_2 1/2(t-t_0)}$. Consequently, our majorizing series does not converge as $t_0 \to -\infty$.

The fact that the majorizing series we constructed does not converge does not mean that the iterates themselves do not converge. Often the iterates (10-4) converge independent of t_0; we just need a way to prove when this happens. The method of this section shows a sufficient condition for the convergence of the iterates that is independent of t_0. Thus, it provides a means for proving existence and uniqueness of steady-state solutions. The proof requires additional restrictions on all of the known functions in Equation (10-3). Thus it is not a generalization of the previous results for all cases. In addition to showing existence and uniqueness, the method gives a uniform upper bound on the response of the system. This bound is useful in worst case system error analysis, and also in circuit problems where a model represents a physical device over a finite range only.

Conversion of the basic equation for a good first iterate. The equation to be solved is still Equation (10-3). The basic method of solution is still successive approximations defining a sequence of functions that hopefully converge to the solution. In the new approach, the first iterate is defined by making a linear approximation to the nonlinear function f. Subsequent iterates bring in the nonlinear effects. The bound on the output is obtained by estimating the effect of the nonlinearity on the output that would result if the system were linear. Thus the first step is to convert (10-3) to a nonlinear integral equation wherein the first iterate is the solution to the linearized problem.

The differential approximation theorem discussed at the end of Section 9-3 above gives a method for getting a linear approximation to the nonlinear function f. Thus we may write

$$f(x(t)) = f(0) + f'(0)\, x(t) + f_1(x(t)) \tag{10-36}$$

where

$$\lim_{x \to 0} \frac{f_l(x)}{x} = 0$$

For the derivations that follow, we shall need to have $f_1(x)$ continuously differentiable. This means that $f''(x)$ must be continuous. Thus we have one more restriction on the nonlinearities of the present section that was not required in the previous section.

Substituting (10-36) into (10-3) gives

$$\begin{aligned} x(t) = g(t) - \int_{t_0}^{t} h(t-\tau)\, f(0)\, d\tau - f'(0) \int_{t_0}^{t} h(t-\tau)\, x(\tau)\, d\tau \\ - \int_{t_0}^{t} h(t-\tau)\, f_1(x(\tau))\, d\tau \end{aligned} \tag{10-37}$$

The term $\int_{t_0}^{t} h(t-\tau)\, f(0)\, d\tau$ is a known time function. It can be combined with $g(t)$ to make up the total forcing function in (10-37). Let

$$g_2(t) = g(t) - \int_{t_0}^{t} h(t-\tau)\, f(0)\, d\tau \tag{10-38}$$

The next term in (10-37) is the number $f'(0)$ multiplied by a convolution of h with x. This term is linear in x. It can be transposed to the left side of the equation to give

$$\begin{aligned} x(t) + f'(0) \int_{t_0}^{t} h(t-\tau)\, x(\tau)\, d\tau \\ = g_2(t) - \int_{t_0}^{t} h(t-\tau)\, f_1(x(\tau))\, d\tau \end{aligned} \tag{10-39}$$

The left side of Equation (10-39) is a linear operator applied to $x(t)$. This operator can be inverted using transforms, provided the inverse operator is stable. To investigate the stability we use a theorem of Paley and Wiener (see Reference 21, pp. 60–61) as follows:

> If $W(s)$ is the transform of an absolutely integrable function $w(t)$ such that $w(t) = 0$ for $t < 0$, then $W(s)/[1 + W(s)]$ is the transform of an absolutely integrable function $v(t)$ for $t > 0$ if and only if $1 + W(s) \neq 0$ for $\mathrm{Re}[s] \geqslant 0$.

In our case $f'(0)\, h(t)$ is the original function. If $h(t)$ is open loop stable, then the first hypothesis of the theorem is satisfied. The second hypothesis is exactly the Nyquist criterion for the linearized system. Thus if this system is closed loop stable, the second hypothesis is satisfied. If $H(s)$ is *PR* the condition is certainly satisfied.

In the transform domain, the operator on the left side of (10-39) is $[1 + f'(0)\, H(s)]$; its inverse is

$$\hat{H}(s) = \frac{1}{1 + f'(0)\, H(s)} = 1 - \frac{f'(0)\, H(s)}{1 + f'(0)\, H(s)} \tag{10-40}$$

By the Paley-Wiener theorem the second term on the right of (10-40) is the transform of an absolutely integrable function $\eta(t)$. The operator in the time domain is

$$\hat{h}(t) = \delta(t) - \eta(t) \tag{10-41}$$

By applying $\hat{h}$ to (10-39) we get

$$\begin{aligned} x(t) = &\int_{t_0}^{t} \hat{h}(t-\tau)\, g_2(\tau)\, d\tau \\ &- \int_{t_0}^{t} \hat{h}(t-\tau) \int_{t_0}^{\tau} h(\tau-\lambda)\, f_1(x(\lambda))\, d\lambda\, d\tau \end{aligned} \tag{10-42}$$

Iteration of the modified equation. This final equation (10-42) has the same basic form as (10-3). A sequence of iterates can be set up as follows:

$$\begin{aligned} x_1(t) &= \int_{t_0}^{t} \hat{h}(t-\tau)\, g_2(\tau)\, d\tau \\ x_2(t) &= x_1(t) - \int_{t_0}^{t} \hat{h}(t-\tau) \int_{t_0}^{\tau} h(\tau-\lambda)\, f_1(x_1(\lambda))\, d\lambda\, d\tau \\ &\vdots \\ x_n(t) &= x_1(t) - \int_{t_0}^{t} \hat{h}(t-\tau) \int_{t_0}^{t} h(\tau-\lambda)\, f_1(x_{n-1}(\lambda))\, d\lambda\, d\tau \end{aligned} \tag{10-43}$$

Since the operator $\hat{h}(t)$ as defined by Equation (10-41) contains an impulse, the procedure used in the previous section to show convergence of the sequence (10-4) does not apply to (10-43). To show conditions that guarantee convergence of the sequence (10-43) we shall use four elementary mathematical concepts. They are (a) the mean-value theorem, (b) the inequality

$$\left| \int_a^b y(x)\, dx \right| \leq \int_a^b |y(x)|\, dx \tag{10-44}$$

(c) the inequality

$$\int_a^b |y(x)|\, |z(x)|\, dx \leq \sup_{a \leq x \leq b} |y(x)| \int_a^b |z(x)|\, dx \tag{10-45}$$

and (d) the geometric series

$$\frac{1}{1-x} = 1 + x + x^2 + \cdots + x^n + \cdots \tag{10-46}$$

when $|x| < 1$.

The first iterate of (10-43) is the variable $x_1(t)$ of the linear feedback system of Figure 10-1. The input $g_1(t)$ takes care of the state at t_0 of the linear system characterized by $h(t)$. The input $f(0)$ accounts for the case where the nonlinear function f in Figure 9-1 does not go through the origin. This system can be analyzed and $x_1(t)$ computed for given

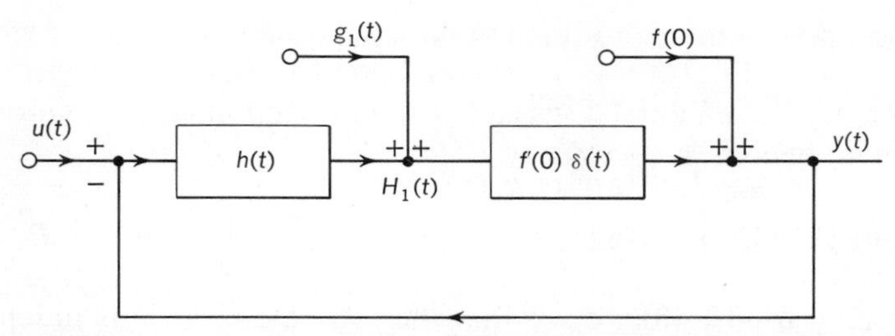

Figure 10-1 Linearized system for computing the first iterate.

$u(t)$ and $g_1(t)$ by the methods of Chapter 4. Furthermore, the stability of the system of Figure 10-1 is the condition that must be satisfied in addition to the absolute integrability of $h(t)$ to guarantee the stability of the operator $\eta(t)$ defined by Equations (10-40) and (10-41). In fact, the operator $\eta(t)$ is exactly the response function of the system of Figure 10-1 when the input is $u(t)$ and the response is $y(t)$. Thus, this linear system diagram and the methods of Chapter 4 give both $x_1(t)$ and the operator $h(t)$ defined by (10-41) for computing subsequent iterates.

Construction of upper bounds for successive iterates. When the system of Figure 10-1 is input-output stable, and $h(t)$ alone is input-output stable, then $x_1(t)$ is bounded whenever $u(t)$ is bounded and the state of the system characterized by $h(t)$ is well-defined at t_0. For $t \geqslant t_0$ let

$$\bar{\mathbf{X}}_1 = \sup |x_1(t)| \tag{10-47}$$

From the iteration formulas (10-43)

$$\begin{aligned} |x_2(t)| &\leqslant \bar{\mathbf{X}}_1 + \left| \int_{t_0}^{t} \hat{h}(t-\tau) \int_{t_0}^{\tau} h(\tau-\lambda)\, f_1(x_1(\lambda))\, d\lambda\, d\tau \right| \\ &\leqslant \bar{\mathbf{X}}_1 + \int_{t_0}^{t} |\hat{h}(t-\tau)| \int_{t_0}^{\tau} |h(\tau-\lambda)|\, |f_1(x_1(\lambda))|\, d\lambda\, d\tau \end{aligned} \tag{10-48}$$

When $x_1(\lambda)$ has a maximum absolute value, $f_1(x_1(\lambda))$ does also because of the assumption that f_1 is continuously differentiable. To show how this bound depends on the bound on $x_1(t)$ we define

$$F_1(\bar{\mathbf{X}}) = \max_{-\bar{\mathbf{X}} \leqslant x \leqslant \bar{\mathbf{X}}} |f'_1(x)| \tag{10-49}$$

Now when $x_1(t)$ is bounded as in (10-47)

$$|f_1(x_1(t))| \leqslant \bar{\mathbf{X}}_1\, F_1\,(\bar{\mathbf{X}}_1) \tag{10-50}$$

With (10-50) the last integral in (10-48) can be estimated by

$$\int_{t_0}^{\tau} |h(\tau-\lambda)|\, |f_1(x_1(\lambda))|\, d\lambda \leqslant \bar{\mathbf{X}}_1\, F_1\,(\bar{\mathbf{X}}_1) \int_{t_0}^{\tau} |h(\tau-\lambda)|\, d\lambda \tag{10-51}$$

Since the upper limit τ in (10-51) takes on values from t_0 to t when the integral appears in (10-48), a courser bound is obtained when τ is set equal to t. By the change of variable used in Section 4-1, Equation (4-5), the estimate becomes

$$\int_{t_0}^{\tau} |h(\tau - \lambda)|\,|f_1(x_1(\lambda))|\,d\lambda \leqslant (\overline{\mathbf{X}}_1)\,F_1\,(\overline{\mathbf{X}}_1) \int_0^{t-t_0} |h(\lambda)|\,d\lambda \tag{10-52}$$

Returning to (10-48) we see that the estimate (10-52) is independent of τ. Thus,

$$|x_2(t)| \leqslant \overline{\mathbf{X}}_1 + \overline{\mathbf{X}}_1\,F_1\,(\overline{\mathbf{X}}_1) \left[\int_0^{t-t_0} |h(\lambda)|\,d\lambda\right] \int_0^{t-t_0} |\hat{h}(\lambda)|\,d\lambda \tag{10-53}$$

Since the last term in (10-53) appears often in the following derivation, it is convenient to define

$$d(\overline{\mathbf{X}}, t) = F_1(\overline{\mathbf{X}}) \left[\int_0^t |h(\lambda)|\,d\lambda\right] \int_0^t |\hat{h}(\lambda)|\,d\lambda \tag{10-54}$$

The function $d(\overline{\mathbf{X}}, t)$ is defined, continuous, positive, and a monotone nondecreasing function of both $\overline{\mathbf{X}}$ and t on $\overline{\mathbf{X}} \geqslant 0$, $t > 0$. Since $\hat{h}$ contains an impulse at $t = 0$, from (10-41)

$$\int_0^t |\hat{h}(\lambda)|\,d\lambda = 1 + \int_0^t |\eta(\lambda)|\,d\lambda \tag{10-55}$$

for $t > 0$. In this case $\eta(\lambda)$ is a function in the usual sense. Since both $\eta(t)$ and $h(t)$ are absolutely integrable, $d(\overline{\mathbf{X}}, t)$ is defined as $t \to \infty$. Let

$$d\,(\overline{\mathbf{X}}) = \lim_{t\to\infty} d\,(\overline{\mathbf{X}}, t) \tag{10-56}$$

In the new notation (10-53) simplifies to

$$|x_2(t)| \leqslant \overline{\mathbf{X}}_1\,[1 + d(\overline{\mathbf{X}}_1, t)] \leqslant \overline{\mathbf{X}}_1\,[1 + d(\overline{\mathbf{X}}_1)] \equiv \overline{\mathbf{X}}_2 \tag{10-57}$$

The general pattern for an upper bound on the nth iterate becomes clear after (10-57) is used in (10-43) to bound $|x_3(t)|$. That is,

$$|x_3(t)| \leqslant \overline{\mathbf{X}}_1 + \int_{t_0}^t |h(t - \tau)| \int_{t_0}^t |h(\tau - \lambda)|\,|f_1(x_2(\lambda))|\,d\lambda\,d\tau \tag{10-58}$$

But for $\lambda \leqslant \tau \leqslant t$

$$\begin{aligned} |f_1(x_2(\lambda))| &\leqslant \max |x_2(t)|\,F_1(\max |x_2(t)|) \\ &\leqslant \overline{\mathbf{X}}_1\,[1 + d(\overline{\mathbf{X}}_1, t)]\,F_1(\overline{\mathbf{X}}_2) \end{aligned} \tag{10-59}$$

Using (10-59) in (10-58) just as (10-50) was used in (10-48) gives

$$|x_3(t)| \leqslant \overline{\mathbf{X}}_1 + \overline{\mathbf{X}}_2\,d(\overline{\mathbf{X}}_2, t) \leqslant \overline{\mathbf{X}}_1 + \overline{\mathbf{X}}_2\,d(\overline{\mathbf{X}}_2) \equiv \overline{\mathbf{X}}_3 \tag{10-60}$$

Continuing this process shows that

$$|x_n(t)| \leqslant \overline{\mathbf{X}}_1 + \mathbf{X}_{n-1}\, d(\overline{\mathbf{X}}_{n-1}) \equiv \overline{\mathbf{X}}_n \tag{10-61}$$

Recapitulation. Thus far we have defined considerable notation, but we still have several more steps to go before obtaining the desired results. Let us pause at this point and review where we are and what we are trying to do. The system problem under consideration is that of Figure 9-1. We are trying to find a set of conditions on the system parameters and the input so that the output is bounded and uniquely determined independent of the state at t_0 as $t_0 \to -\infty$. Furthermore, we should like an easily computed formula that gives an upper bound on the system response as a function of the maximum value of the input to the system.

Toward this goal we have defined a sequence of iterates, (10-43). Below we shall exhibit a set of conditions under which these iterates converge to the desired system response. When these conditions are satisfied the response is unique as $t_0 \to -\infty$. In (10-61) we have a recursive formula for a bound on the iterates. We shall show that when the conditions for the convergence of the iterates are satisfied, the sequence of numbers $\overline{\mathbf{X}}_n$ converges and that the limit of this sequence is a bound on the system response. Furthermore, we shall show that this limit can be estimated by a formula that gives an upper bound on the response as a function of the maximum value of the input.

Demonstration of convergence. To show convergence of the iterates we first convert the sequence of functions (10-43) to an infinite series problem by investigating the series $\Sigma_{k=1}^{\infty}\ D_k(t)$ where

$$\begin{aligned} D_k(t) &= x_k(t) - x_{k-1}(t) \quad k \geqslant 2 \\ D_1(t) &= x_1(t) \end{aligned} \tag{10-62}$$

Then for $k > 2$

$$D_k(t) = -\int_{t_0}^{t} \hat{h}(t-\tau) \int_{t_0}^{\tau} h(\tau-\lambda)\ [f_1(x_{k-1}(\lambda)) - f_1(x_{k-2}(\lambda))]\ d\lambda\ d\tau \tag{10-63}$$

This term can be estimated by use of the mean-value theorem and the usual absolute value estimates for integrals. This estimate will relate a bound on D_k to the quantities already defined in the bounds $\overline{\mathbf{X}}_n$.

In order to get the correct terms from (10-61) into the bound on the D_k we must show the following:

Lemma: $\qquad \mathbf{X}_n \geqslant \overline{\mathbf{X}}_m$ when $n > m$ $\qquad$ (10-64)

Proof: The proof is by induction. Clearly from (10-57) $\overline{\mathbf{X}}_2 \geqslant \overline{\mathbf{X}}_1$ because of the properties of $d(\overline{\mathbf{X}}, t)$ given after the definition (10-54). For the induction step we assume $\overline{\mathbf{X}}_{n-1} \geqslant \overline{\mathbf{X}}_{n-2} \geqslant \cdots \geqslant \overline{\mathbf{X}}_1$. Then we must show from formula (10-61) that $\overline{\mathbf{X}}_n \geqslant \overline{\mathbf{X}}_{n-1}$. Since all terms are positive we can examine $\overline{\mathbf{X}}_n - \overline{\mathbf{X}}_{n-1}$. If this quantity is positive, the Lemma is proved. Using (10-61) for $\overline{\mathbf{X}}_n$ and $\overline{\mathbf{X}}_{n-1}$

$$\overline{\mathbf{X}}_n - \overline{\mathbf{X}}_{n-1} = \overline{\mathbf{X}}_{n-1}\, d(\overline{\mathbf{X}}_{n-1}) - \overline{\mathbf{X}}_{n-2}\, d(\overline{\mathbf{X}}_{n-2}) \tag{10-65}$$

Since $d(\overline{\mathbf{X}})$ is monotone nondecreasing, $d(\overline{\mathbf{X}}_{n-1}) \geqslant d(\overline{\mathbf{X}}_{n-2})$ when $\overline{\mathbf{X}}_{n-1} \geqslant \overline{\mathbf{X}}_{n-2}$. Thus (10-65) is positive and the Lemma is proved.

Returning to (10-63), applying the mean-value theorem to $[f_1(x_{k-1}(\lambda)) - f_1(x_{k-2}(\lambda))]$, using the Lemma and the various definitions above gives

$$|D_k(t)| \leqslant \int_{t_0}^{t} |\hat{h}(t-\tau)| \int_{t_0}^{\tau} |h(t-\tau)|\; F_1(\overline{\mathbf{X}}_{k-1})\; |D_{k-1}(\tau)|\; d\tau \tag{10-66}$$

Using (10-66) recursively for $k \geqslant 2$ gives

$$\begin{aligned}
|D_1(t)| &\leqslant \overline{\mathbf{X}}_1 \\
|D_2(t)| &\leqslant d(\overline{\mathbf{X}}_1)\, \overline{\mathbf{X}}_1 \\
|D_3(t)| &\leqslant d(\overline{\mathbf{X}}_1)\, d(\overline{\mathbf{X}}_2)\, \overline{\mathbf{X}}_1 \\
&\vdots \\
|D_k(t)| &\leqslant d(\overline{\mathbf{X}}_1)\, d(\overline{\mathbf{X}}_2) \cdots d(\overline{\mathbf{X}}_{k-1})\, \overline{\mathbf{X}}_1
\end{aligned} \tag{10-67}$$

By the Lemma and the properties of $d(\overline{\mathbf{X}})$, each $d(\overline{\mathbf{X}}_n)$ can be majorized by $d(\overline{\mathbf{X}}_k)$ when $n < k$. Thus,

$$|D_k(t)| \leqslant [d(\overline{\mathbf{X}}_{k-1})]^{k-1}\, \overline{\mathbf{X}}_1 \tag{10-68}$$

If the $\overline{\mathbf{X}}_k$ are bound by some $\overline{\mathbf{X}}_0$ and $d(\overline{\mathbf{X}}_0) < 1$, the ratio test shows that the series $\Sigma_{k=1}^{\infty} D_k(t)$ converges uniformly and absolutely for all t. Then

$$|D_k(t)| \leqslant [d(\overline{\mathbf{X}}_0)]^{k-1}\, \overline{\mathbf{X}}_1$$

and

$$\left| \sum_{k=1}^{\infty} D_k(t) \right| \leqslant \frac{\overline{\mathbf{X}}_1}{1 - d(\overline{\mathbf{X}}_0)} \tag{10-69}$$

The right side of (10-69) is also an upper bound on every $\overline{\mathbf{X}}_k$. Thus a quantity equal to the right side is appropriate for $\overline{\mathbf{X}}_0$. In other words, we can find $\overline{\mathbf{X}}_0$ by the implicit equation

$$\overline{\mathbf{X}}_0 = \frac{\overline{\mathbf{X}}_1}{1 - d(\overline{\mathbf{X}}_0)} \tag{10-70}$$

If we can find an $\overline{\mathbf{X}}_0$ satisfying (10-70) and such that $d(\overline{\mathbf{X}}_0) < 1$, then we have the conditions for convergence of the iterates. We also have an upper bound on the total response $x(t)$ for the system.

Step-by-step procedure for bounding the response. By combining (10-70) with the formula for $d(\overline{\mathbf{X}})$ in terms of system quantities, we can find conditions on these quantities such that the equations are satisfied for a $d(\overline{\mathbf{X}}_0) < 1$. Then starting from these conditions we can show that the iteration procedure can be used to construct a set of $\overline{\mathbf{X}}_k$ that are suitably bounded, and the sequence $\overline{\mathbf{X}}_n$ is bounded by $\overline{\mathbf{X}}_0$. Let us state the steps specifically in symbols and then consider a numerical example that shows the computational steps.

From (10-54) and (10-56)

$$d(\overline{\mathbf{X}}) = F_1(\overline{\mathbf{X}}) \left[\int_0^\infty |h(\lambda)| \, d\lambda \right] \int_0^\infty |\hat{h}(\lambda)| \, d\lambda \tag{10-71}$$

Recall that $F_1(\overline{\mathbf{X}})$ is defined by (10-49) to be the maximum of the derivative of f_1 defined by (10-36) in the interval $-\overline{\mathbf{X}} \leqslant x \leqslant \overline{\mathbf{X}}$. The assumption that the original nonlinear function f was twice continuously differentiable implies that f_1 is continuously differentiable. Furthermore, the limit condition given under (10-36) shows that $f_1(0) = 0$ and $f'_1(0) = 0$. Thus, $F_1(\overline{\mathbf{X}})$ is a monotone, nondecreasing function and $F_1(0) = 0$. Applying this fact to (10-71) shows that $d(0) = 0$ and d is a continuous function. Thus $d(\overline{\mathbf{X}})$ is small when $\overline{\mathbf{X}}$ is small.

The requirement $d(\overline{\mathbf{X}}_0) < 1$ and formulas (10-70) and (10-71) can be satisfied if the following equation has a solution $d < 1$. It is

$$d = F_1\left(\frac{\overline{\mathbf{X}}_1}{1-d}\right) \left[\int_0^\infty |h(\lambda)| \, d\lambda \right] \int_0^\infty |\hat{h}(\lambda)| \, d\lambda \tag{10-72}$$

For any fixed value of $d < 1$ the right side of (10-72) can be made as small as we please by making $\overline{\mathbf{X}}_1$ small. Thus, except in the very special case where $F_1(\overline{\mathbf{X}})$ is very small for all $\overline{\mathbf{X}}$ so that the right side cannot be as large as one, for any $d < 1$ there is an $\overline{\mathbf{X}}_1$ that satisfies (10-72). With this $\overline{\mathbf{X}}_1$ and d, the other conditions will be satisfied with

$$\overline{\mathbf{X}}_0 = \frac{\overline{\mathbf{X}}_1}{1-d} \tag{10-73}$$

The procedure to show that the iterates do converge when the input is not too large is as follows:

1. Select a value of $d < 1$.
2. Use formula (10-72) to find an $\overline{\mathbf{X}}_1$ that satisfies the equation for the selected value of d.
3. Use $\overline{\mathbf{X}}_1$ in the recursion formulas (10-61) and (10-67). In these formulas $d(\overline{\mathbf{X}}_1), d(\overline{\mathbf{X}}_2), \cdots, d(\overline{\mathbf{X}}_k), \cdots$ will all be less than d and all $\mathbf{X}_k$ will be bounded by $\dfrac{\overline{\mathbf{X}}_1}{1-d}$.
4. Compare the series generated in step 4 for bounding the $|D_k(t)|$

to the geometric series $(1 + d + d^2 + \cdots + d^k + \cdots)\,\overline{\mathbf{X}}_1$. Since each term of the series to be tested is less than the corresponding term in a series that is known to converge, the series in question converges. Thus, the sequence of iterates converges uniformly and absolutely when the input is restricted so that the first iterate $x_1(t)$ is bounded by $\overline{\mathbf{X}}_1$.

5. The first iterate $\mathbf{x}_1(t)$ is the response of a linear system. When the functional form of the input is given, the maximum of the output is linear in the maximum of the input. Then $\overline{\mathbf{X}}_1$ can be related to a range of allowed inputs.

An illustrative example. Those steps of the above procedure that need to be computed to obtain the input-output bound are best illustrated by an example. For this case we choose the various functions simple enough for computation by slide rule. Thus the linear system represented by $h(t)$ in Figure 9-1 is a second-order system with damped sinusoidal response. The nonlinearity is chosen as a small square law correction on an otherwise linear function. The input is chosen as a steady-state sinusoid at the resonant frequency of the linearized system of Figure 10-1. Since we assume the steady state, $g_1(t)$, the state at t_0 term in Figure 10-1, is zero. Also, for simplicity we take a nonlinear function that goes through the origin. Thus, $f(0)$ in Figure 10-1 is zero.

Specifically the terms in Figure 9-1 are

$$u(t) = A \cos \omega_0 t \tag{10-74}$$

$$h(t) = \begin{cases} 0 & \text{for } t < 0 \\ 2e^{-t} \cos 10t & \text{for } t \geqslant 0 \end{cases} \tag{10-75}$$

$$f(x) = 5(1 + 0.1x)\,x \tag{10-76}$$

Here A is left free since it is the bound on the input that must be related to the bound on the output. The input frequency ω_0 is determined below to be the resonant frequency of the system of Figure 10-1 or the imaginary part of the pole of $\hat{H}(s)$ defined by (10-40).

The first quantity that must be computed is $f_1(x)$ from (10-36).

$$\begin{aligned} f_1(x) &= f(x) - f(0) - f'(0)\,x \\ &= 5(1 + 0.1x)\,x - 0 - 5x = \tfrac{1}{2}x^2 \end{aligned} \tag{10-77}$$

Next we transform $h(t)$ to get

$$H(s) = \frac{1}{s + 1 + j10} + \frac{1}{s + 1 - j10} = \frac{2(s + 1)}{(s + 1)^2 + 10} \tag{10-78}$$

Then from (10-40)

$$\hat{H}(s) = 1 - \frac{5\dfrac{2(s+1)}{(s+1)^2+10}}{1+5\dfrac{2(s+1)}{(s+1)^2+10}} = 1 - 10\,\frac{s+1}{s^2+12s+12} \tag{10-79}$$

Factoring the denominator gives

$$\omega_0 = 9.22 \tag{10-80}$$

Inverse transforming gives

$$\hat{h}(t) = \delta(t) - 11.40\, e^{-6t} \cos\,(9.22t + 0.496) \tag{10-81}$$

Since this example has both the zero-input term and the $f(0)$ term equal to zero, the first iterate can be found by conventional steady-state analysis. That is,

$$\begin{aligned} g_2(t) = g(t) &= \int_{t_0}^{t} h(t-\tau)\, u(\tau)\, d\tau \\ &= A|H(j\omega_0)|\, \cos\,(\omega_0 t + \underline{/H(j\omega_0)}) \end{aligned} \tag{10-82}$$

Similarly for (10-43)

$$\begin{aligned} x_1(t) &= \int_{-\infty}^{t} \hat{h}(t-\tau)\, g_2(\tau)\, d\tau \\ &= A|\hat{H}(j\omega_0)|\, |H(j\omega_0)|\, \cos\,(\omega_0 t + \underline{/H(j\omega_0)} + \underline{/\hat{H}(j\omega_0)}) \end{aligned} \tag{10-83}$$

The only computation we need to carry out is that for $\overline{\mathbf{X}}_1$, the upper bound on $x_1(t)$. Now

$$\overline{\mathbf{X}}_1 = A|H(j\omega_0)|\, |\hat{H}(j\omega_0)| = 1.34A \tag{10-84}$$

The next computation is $F_1(\overline{\mathbf{X}})$ from (10-49). Since $f_1(x)$ is so simple

$$F_1(\overline{\mathbf{X}}) = \overline{\mathbf{X}} \tag{10-85}$$

With this formula we can compute $d(\overline{\mathbf{X}})$ from (10-71). This requires computation of the two constants $\int_0^\infty |h(\lambda)|\, d\lambda$ and $\int_0^\infty |\hat{h}(\lambda)|\, d\lambda$. Since both $h(t)$ and $\hat{h}(t)$ are periodic functions multiplied by decaying exponentials the integrals are readily computed. Each cycle is merely $e^{-\alpha T}$ times the previous cycle. Since $e^{-\alpha T}$ is less than one, the infinite series of areas under succeeding cycles is $1/(1-e^{-\alpha T})$ times the area of one cycle. Specifically, from (10-75)

$$\begin{aligned} \int_0^\infty |h(\lambda)|\, d\lambda &= \int_0^\infty 2e^{-\lambda}\, |\cos 10\lambda|\, d\lambda \\ &= \frac{2}{1-e^{-\pi/10}} \int_0^{\pi/10} e^{-\lambda} \cos 10\lambda\, d\lambda \\ &= 1.304 \end{aligned} \tag{10-86}$$

Since $\hat{h}(t)$ changes sign in the first half period, it takes a little more effort to obtain the integral of its absolute value. From (10-82)

$$\begin{aligned}\int_0^\infty |\hat{h}(\lambda)|\,d\lambda &= 1 + \int_0^\infty 11.4e^{-6\lambda}\,|\cos(9.22\lambda + 0.496)|\,d\lambda \\ &= 1 + \frac{11.4}{1 - e^{-6(0.341)}}\left[\int_0^{0.1167} e^{-6\lambda}\cos(9.22\lambda + 0.496)\,d\lambda \right. \\ &\qquad \left. - \int_{0.1167}^{0.341} e^{-6\lambda}\cos(9.22\lambda + 0.496)\,d\lambda\right] \\ &= 2.556\end{aligned} \tag{10-87}$$

Now from (10-71)

$$d(\bar{\mathbf{X}}) = (2.556)(1.304)\,\bar{\mathbf{X}} = 3.33\,\bar{\mathbf{X}} \tag{10-88}$$

In (10-72) the numbers give

$$d = \frac{\bar{\mathbf{X}}_1}{1 - d}(3.33) \tag{10-89}$$

or

$$d^2 - d + 3.33\,\bar{\mathbf{X}}_1 = 0 \tag{10-90}$$

The roots of this quadratic are

$$d = \tfrac{1}{2} \pm \sqrt{\tfrac{1}{4} - 3.33\,\bar{\mathbf{X}}_1} \tag{10-91}$$

The largest $\bar{\mathbf{X}}_1$ for which there is a real $d < 1$ occurs when the radical is zero. Thus the series will converge so long as

$$\bar{\mathbf{X}}_1 < 0.075$$

To relate $\bar{\mathbf{X}}_1$ to the input amplitude A we use (10-84). Thus the iterates converge so long as A is less than 0.056.

Now to compute an output bound for a given input we must first be sure that A is less than 0.056. Then we use (10-84) to compute $\bar{\mathbf{X}}_1$. The actual response $x(t)$ is bounded by $\bar{\mathbf{X}}_1/(1 - d)$, where d is given by (10-91) with the minus sign. Thus

$$\bar{\mathbf{X}} < \frac{1.34A}{\tfrac{1}{2} + \sqrt{\tfrac{1}{4} - 4.46A}} \tag{10-92}$$

Examining (10-92) we see that for very small A the denominator approaches one and the bound approaches $\bar{\mathbf{X}}_1$ as it should.

Uniqueness of the solution. The derivation above only went to the point of showing convergence of the iterates. In the example we assumed that the iterates do converge to the solution and that the solution is unique. These assumptions must be shown to be valid. The proof that $x(t)$ is a solution as outlined at the end of the previous section relied on the fact that the series of D_k converged. In this case we still have a convergent series of D_k, so the same approach is valid. For uniqueness the procedure of the previous section also applies. The difference is that as the steps corresponding to those of (10-31), (10-32), and (10-33) are repeated, the right sides of the inequalities are successive terms in the sequence (10-67). These terms approach zero as k approaches infinity provided the two solutions both are bounded so that $d(\overline{\mathbf{X}})$ and $d(\overline{\overline{\mathbf{X}}})$ are less than one. Thus, we can only prove that there is a unique solution with the bound $\overline{\mathbf{X}}_0$ of (10-70). From the results of this section alone there could be a second solution that is not bounded by $\overline{\mathbf{X}}_0$.

By combining the results of this section with those of the previous section we can show a class of problems that do have unique steady-state solutions. This class consists of those systems for which the procedure of this section gives a convergent series independent of t_0 when both transient and steady-state terms are considered in $g_2(t)$ of (10-38). For finite t_0 the results of the previous section show that the unique response is suitably bounded. Now should there be a second solution to the equation when t_0 is $-\infty$, we know that it cannot be generated by setting up a real physical problem and waiting for a steady state to appear.

10-3 Stability Using both Integral and Differential Equations

In Section 9-3 the concept of Lyapunov stability was presented. The problem at the end of that section was as follows:

Given a system characterized by

$$\dot{\mathbf{x}} = \mathbf{F}(\mathbf{x}, t) \tag{10-93}$$

an initial condition $\mathbf{x}_0(t_0)$, and the corresponding response $\mathbf{x}_0(t)$, show the region of initial conditions $\mathbf{x}_1(t_0)$ for which $\mathbf{x}_0(t)$ is stable in the sense of Lyapunov.

In the first-order case (9-30) the problem becomes a study of the stability of the origin of (9-34) for the initial condition

$$\Delta(t_0) = x_1(t_0) - x_0(t_0) \tag{10-94}$$

The problem can be converted to an integral equation problem by regrouping (9-34) with linear terms on the left and the nonlinear term on the right, and then inverting the linear operator. The regrouping gives

$$\dot{\Delta} - \frac{\partial f(x_0, t)}{\partial} \Delta = R(\Delta) \tag{10-95}$$

The problems of inverting the operator on the left of (10-95) is exactly the same as the problem of solving the linear variable coefficient equation (3-23). By using the results of Section 3-2 as given by (3-31) we get

$$\Delta(t) = \int_{t_0}^{t} \exp\left[\int_{t_0}^{t} \frac{\partial f(x_0(v), v)}{\partial x}\, dv - \int_{t_0}^{\tau} \frac{\partial f(x_0(v), v)}{\partial x}\, dv\right] R(\Delta(\tau))\, d\tau$$
$$+ \Delta(t_0) \exp\left[\int_{t_0}^{t} \frac{\partial (x_0(\tau), \tau)}{\partial x}\, d\tau\right] \tag{10-96}$$

Since $\Delta(t_0)$ is a number, the last term in (10-96) is a known time function—call it $g(t)$. The other term on the right of (10-96) has the form $\int_{t_0}^{t} h(t, \tau)\, R(\Delta(\tau))\, d\tau$, where $h(t, \tau)$ is the response function of a linear, time-variant system. In the new notation (10-96) becomes

$$\Delta(t) = g(t) + \int_{t_0}^{t} h(t, \tau)\, R(\Delta(\tau))\, d\tau \tag{10-97}$$

This equation is similar to (10-24) in all the ways required for the methods of Section 10-2 to apply with only minor modification. An input-output result on (10-97) similar to the result of Section 10-2 is exactly what is needed to show a range of Lyapunov stability for the response $x_0(t)$ of the first-order system of the form of (10-94).

The properties of the various functions on the right of (10-97) that must be true for the previous results to apply are

1. The maximum value of $|g(t)|$ must be proportional to the initial condition $|\Delta(t_0)|$.

2. The response function $h(t, \tau)$ must be absolutely integrable for all t. That is, there must be an H_1 such that

$$\int_{t_0}^{t} |h(t, \tau)|\, d\tau \leq H_1 \qquad \text{for all } t$$

3. The nonlinear remainder function R must be continuously differentiable and $R(0) = 0$.

The first and second properties are assured if the linear, time-variant system is input-output stable. The inequality in the second property is the condition for input-output stability for linear, time-variant systems corresponding to that for time-invariant systems as given in the theorem of Section 8-3. If that inequality is satisfied, $|g(t)|$ satisfies property 1. Property 3 will be satisfied if the original nonlinear function in (10-93) has a continuous second partial with respect to x.

For higher-order systems the equation for the difference between two solutions corresponding to (10-95) is a regrouping of the vector equation (9-38). That is

$$\dot{\Delta} - \frac{\partial \mathbf{F}}{\partial \mathbf{x}} \Delta = \mathbf{R}(\Delta) \tag{10-98}$$

The first step in setting up this problem for the methods of the previous section is to find the vector response function for the linear, time-variant system characterized by the operator on the left of (10-98). Since there is no general method for getting such a solution in the form of a formula suitable for applying the results available, there is no point in proceeding further with a detailed discussion. It suffices to state that if this linear system is input-output stable, then the method outlined for the first-order system can be applied to the problem for small perturbations of the initial conditions — $\|\Delta(t_0)\|$ small.

Although we cannot proceed with the method to get a δ for each ϵ in the definition of Lyapunov stability, the above results do have considerable engineering significance as follows: Most nonlinear systems cannot be solved analytically. Thus the engineer analyzes the system by getting a numerical solution on a digital computer. The results of each computation is one solution for one set of initial conditions. This result is not of much value unless the solution is more or less right even if the initial conditions on the actual system are not exactly those fed to the computer. Now the problem of investigating the validity of a solution for small perturbations of initial conditions has been reduced to the problems of investigating the input-output stability of a linear, time-variant system. This stability problem is often easier than the investigation of the nonlinear system for all types of small perturbations of initial conditions.

A simple example. As an example of the steps of the stability criterion just outlined let us consider a simple nonlinear equation for which there is an explicit formula for the response. That is, we apply the method to an exact differential equation of first order (see Reference 11, pp. 44–51). One such equation for which the formulas are simple is

$$\dot{x} = -\frac{2x - 3t}{2(2x + t)} \tag{10-99}$$

For this equation the right side fits the notation above

$$f(x, t) = -\frac{2x - 3t}{2(2x + t)} \tag{10-100}$$

The solution to this equation can be written in the form

$$x(t) = -\frac{t}{2} \pm \sqrt{t^2 + C^2} \tag{10-101}$$

where C^2 is a constant that sets the formula to a given initial state. For simplicity we take the initial state at $t_0 = 0$.

If $x(0)$ is positive, the plus sign in (10-101) is appropriate. Then $C = x(0)$. If $x(0)$ is negative, the minus sign in (10-101) is appropriate. Then $C = -x(0)$. In either case C is a positive constant. In either case the system is not stable. If $x(0) > 0$ the response grows linearly in t. If $x(0) < 0$ the response goes to $-\infty$ linearly as $t \to \infty$. Nevertheless, given a solution, any other solution starting with the same initial sign does approach the first solution.

The definition of stability in the sense of Lyapunov applies to a particular solution $x_0(t)$. To be specific take $x_0(0) = C_0 > 0$ and then

$$x_0(t) = -\frac{t}{2} + \sqrt{t^2 + C_0^2} \tag{10-102}$$

We then compare this solution with a second solution $x_1(t)$. If $x_1(0) = C_1 > 0$, then

$$\Delta(t) = x_1(t) - x_0(t) = \sqrt{t^2 + C_1^2} - \sqrt{t^2 + C_0^2} \tag{10-103}$$

Expanding the square roots in power series shows that $\Delta(t) \to 0$ as $t \to \infty$. Specifically,

$$\sqrt{t^2 + C^2} = t + \frac{1}{2}\frac{C^2}{t} + 0\left(\frac{1}{t^3}\right) \tag{10-104}$$

where $0\ (1/t^3)$ means terms that go asymptotically such as

$$\lim_{t \to \infty} t^3\, 0\left(\frac{1}{t^3}\right) = \text{a constant.}$$

Thus,

$$\Delta(t) = \frac{1}{2t}(C_1^2 - C_0^2) + 0\left(\frac{1}{t^3}\right) \tag{10-105}$$

As $t \to \infty, \Delta(t) \to 0$ provided $x_1(0) > 0$ so that formula (10-103) is valid. If $x_1(0) < 0$ then $x_1(t) \to -\infty$ while $x_0(t) \to +\infty$ as $t \to \infty$. In the definition of Lyapunov stability (9-24) this means that for given x_0, $\delta < x_0(0)$ for any ϵ.

Let us now apply the formulas of this section to (10-99) with given solution (10-102). Differentiating (10-100) gives

$$\frac{\partial f}{\partial x} = -\frac{4t}{(2x + t)^2} \tag{10-106}$$

Substituting (10-102) into (10-106) gives the coefficient needed for (10-95). It is

$$\frac{\partial f}{\partial x}(x_0, t) = -\frac{t}{t^2 + C^2} \tag{10-107}$$

The other term needed for (10-95) is $R(\Delta)$. This is defined in (9-32). It is

$$\begin{aligned} R(\Delta) &= f(x_1, t) - f(x_0, t) - \frac{\partial f(x_0, t)}{\partial x}\Delta \\ &= -\frac{2x_1 - 3t}{2(2x_1 + t)} - \left[\frac{t}{\sqrt{t^2 + C_0^2}} - \frac{1}{2}\right] + \frac{t}{t^2 + C_0^2}\Delta \end{aligned} \tag{10-108}$$

The right side of (10-108) should be in terms of Δ and t. The x_1 terms can be eliminated since

$$\Delta = x_1 - x_0 = x_1 + \tfrac{1}{2}t - \sqrt{t^2 + C_0^2} \tag{10-109}$$

Solving for x_1 gives

$$x_1 = \Delta - \tfrac{1}{2}t + \sqrt{t^2 + C_0^2} \tag{10-110}$$

Substituting in (10-108) gives

$$R(\Delta) = \frac{t\Delta^2}{(t^2 + C_0^2)(\Delta + \sqrt{t^2 + C_0^2})} \tag{10-111}$$

Clearly $R(\Delta)$ is bounded for all $t \geqslant 0$ and

$$\lim_{\Delta \to 0} \frac{\text{max for all } t \text{ of } R(\Delta)}{\Delta} = 0 \tag{10-112}$$

In order to estimate the range of Lyapunov stability of the solution $x_0(t)$ we must compute $g(t)$ and $h(t, \tau)$ in (10-97). These quantities are given explicitly in (10-96). The first step is computation of the integral for the exponential function. In the present case with $t_0 = 0$, (10-107) in the integral gives

$$\begin{aligned}\int_0^t \frac{\partial f(x_0(v), v)}{\partial x}\, dv &= -\int_0^t \frac{v}{v^2 + C_0^2}\, dv \\ &= -\tfrac{1}{2}\ln(t^2 + C_0^2) + \tfrac{1}{2}\ln C_0^2 = -\ln \frac{\sqrt{t^2 + C_0^2}}{C_0}\end{aligned} \tag{10-113}$$

Now

$$g(t) = \frac{\Delta(0)\, C_0}{\sqrt{t^2 + C_0^2}} \tag{10-114}$$

Furthermore

$$h(t, \tau) = \exp\left[-\ln \frac{\sqrt{t^2 + C_0^2}}{C_0} + \ln \frac{\sqrt{\tau^2 + C_0^2}}{C_0}\right] = \sqrt{\frac{\tau^2 + C_0^2}{t^2 + C_0^2}} \tag{10-115}$$

Substituting Equations (10-111), (10-114), and (10-115) into (10-97) gives the integral equation for $\Delta(t)$ as

$$\Delta(t) = \frac{\Delta(0) C_0}{\sqrt{t^2 + C_0^2}} + \int_0^t \sqrt{\frac{\tau^2 + C_0^2}{t^2 + C_0^2}} \left(\frac{\tau \Delta^2(\tau)}{(\tau^2 + C_0^2)(\Delta(\tau) + \sqrt{\tau^2 + C_0^2})}\right) d\tau \tag{10-116}$$

The integral equation (10-116) must be compared with the estimates of the previous section to get an estimate on max $\Delta(t)$ given $\Delta(0)$. We could proceed estimating both R and h from (10-114) and (10-115). However, since there is some cancellation in the integral (10-116) we can take advantage of certain simplifications before making the estimates. Let us rewrite (10-116) as

$$\Delta(t) = \frac{\Delta(0)C_0}{\sqrt{t^2+C_0^2}} + \frac{1}{\sqrt{\tau^2+C_0^2}}\int_0^t \left(\frac{\tau}{\sqrt{\tau^2+C_0^2}}\right)\left(\frac{\Delta^2(\tau)}{\Delta(\tau)+\sqrt{\tau^2+C_0^2}}\right) d\tau \tag{10-117}$$

Comparing (10-117) to the integral equation (10-42) of the previous section we see that the quantities correspond as follows

$$x_1(t) \sim \frac{\Delta(0)C_0}{\sqrt{t^2+C_0^2}} \tag{10-118}$$

Then

$$\overline{\mathbf{X}}_1 \sim \Delta(0) \tag{10-119}$$

Next

$$f_1(x) \sim \frac{\Delta^2}{\Delta+\sqrt{t^2+C_0^2}} \tag{10-120}$$

Then

$$\begin{aligned} F_1(\overline{\mathbf{X}}) &\sim \max_{\substack{-\overline{\Delta}\leq\Delta<\overline{\Delta}\\ t\geq 0}} \left|\frac{d}{d\Delta}\left[\frac{\Delta^2}{\Delta+\sqrt{t^2+C_0^2}}\right]\right| \\ &= \max_{\substack{-\overline{\Delta}<\Delta<\overline{\Delta}\\ t>0}} \left|\frac{\Delta\left(\Delta+2\sqrt{t^2+C_0^2}\right)}{\left(\Delta+\sqrt{t^2+C_0^2}\right)^2}\right| \end{aligned} \tag{10-121}$$

This term needs to be examined in more detail. This examination is done below after we show the final required correspondence to (10-54). It is

$$\frac{d(\mathbf{X},t)}{F_1(\mathbf{X})} \sim \frac{1}{\sqrt{t^2+C_0^2}}\int_0^t \frac{\tau}{\sqrt{\tau^2+C_0^2}}\, d\tau = 1 - \frac{C_0}{\sqrt{t^2+C_0^2}} \tag{10-122}$$

Then

$$\frac{d(\mathbf{X})}{F_1(\mathbf{X})} \sim 1 \tag{10-123}$$

The above formulas are all very simple except (10-121), the estimate corresponding to F_1. One fact is immediately clear from the formula.

That is, $\bar{\Delta}$ must be less than C_0[4] because otherwise the denominator goes to zero for $\Delta = C_0, t = 0$. A check of the partial derivatives of the last expression in (10-121) shows that the quantity in the absolute value signs is monotonic in both t and Δ for all Δ and $t \geqslant 0$. In t, the maximum occurs at $t = 0$. For Δ we must check $\Delta = \bar{\Delta}$ and $\Delta = -\bar{\Delta}$. The largest absolute value occurs when $\Delta = -\bar{\Delta}$. Let us call this maximum $F_1(\bar{\Delta})$. It is

$$F_1(\bar{\Delta}) = \frac{\bar{\Delta}(2C_0 - \bar{\Delta})}{(C_0 - \bar{\Delta})^2} \tag{10-124}$$

Clearly $F_1(\bar{\Delta})$ is monotonic in $\bar{\Delta}$ for $0 \leqslant \bar{\Delta} \leqslant C_0$.

To show a range of $\Delta(0)$ for which the system is stable in the sense of Lyapunov we must relate d and $\Delta(0)$. We must find the maximum $\Delta(0)$ for which there is a $d < 1$. This can be done through manipulation of (10-119), (10-123), (10-124), and (10-72). In terms of present quantities (10-72) becomes

$$d = \frac{\dfrac{\Delta(0)}{1-d}\left(2C_0 - \dfrac{\Delta(0)}{1-d}\right)}{\left(C_0 - \dfrac{\Delta(0)}{1-d}\right)^2} \tag{10-125}$$

This equation gives the relationship between $\Delta(0)$ and d for specified C_o. Let us choose $C_0 = 1$ to be specific.

Now (10-125) is cubic in d and quadratic in $\Delta(0)$. Of the various solutions, the usable ones must be such that $0 \leqslant \Delta(0) < C_0 = 1$. This restriction arises because (10-124) is valid only for $0 \leqslant \bar{\Delta} < C_0 = 1$ and $\bar{\Delta} = \Delta(0)/(1-d)$. Furthermore, $0 < d < 1$, because the iteration procedure of the previous section requires such bounds on d. Since quadratics are easier to solve than cubics, let us rewrite (10-125) as a quadratic in $\Delta(0)$. The result is

$$(d+1)\ \Delta^2(0) + 2(d^2-1)\ \Delta(0) + (d^3 - 2d^2 + d) = 0 \tag{10-126}$$

Solving for $\Delta(0)$ by the quadratic formula gives

$$\Delta(0) = \frac{1-d}{1+d}\,[1 + d \pm \sqrt{1+d}] \tag{10-127}$$

Since $d < 1$ and $\bar{\Delta} = \Delta(0)/(1-d) < 1$, the minus sign is appropriate.

The problem now reduces to finding the largest value of $\Delta(0)$ in (10-127) with the minus sign, while maintaining the restriction $d < 1$. First we note that when $d = 0$ and when $d = 1$, $\Delta(0) = 0$. For $0 < d < 1$, $\Delta(0)$ is positive. Differentiating (10-127) with respect to d and setting this derivative to zero locates the maximum. The formula is

[4] Recall that we are investigating a solution with $x_0(0) = C_0 > 0$.

$$3 + d - 2(1 + d)\sqrt{1 + d} = 0 \tag{10-128}$$

Since (10-128) is a cubic in d, it is not readily solved. For small d, the left side of (10-128) is positive and for larger d it is negative. By trial and error we find the zero at $d = 0.43$. Substituting this value in (10-127) with the minus sign gives

$$\Delta(0) = 0.0916 \tag{10-129}$$

With (10-129) for $\Delta(0)$ and the corresponding value of d we have the estimate of the maximum difference between the given solution $x_0(t)$ and some other solution $x_1(t)$ such that $|x_1(t) - x_0(t)| < \bar{\Delta}$ as

$$\bar{\Delta} \leqslant \frac{\Delta(0)}{1-d} = 0.161 \tag{10-130}$$

In terms of the definition of stability in the sense of Lyapunov the above computations show that for $\epsilon = 0.161$, $\delta = 0.0916$ is satisfactory.

PROBLEMS

10-1 Consider the circuit of Figure 9-1 with

$$u(t) = A \cos 9.22\, t$$

$$h(t) = \begin{cases} 0 & \text{for } t < 0 \\ 2e^{-t} \cos 10t & \text{for } t \geqslant 0 \end{cases}$$

$$f(x) = 5(1 + 0.1\, x^2)\, x$$

Find the range of values of A for which the procedure of Section 10-2 guarantees input-output stability. Also find a bound on the output similar to (10-92).

10-2 Consider the circuit of Figure 9-1 with the nonlinearity having the form

$$f(x) = K(1 + 0.1x)\, x$$

Let the linear system be third order and select K and $h(t)$ so that the resulting $h(t) = K_1e^{-t} + K_ee^{-2t} + K_3e^{-3t}$ for $t \geqslant 0$. Let

$$u(t) = A \cos 2t$$

Find the range of A for which the procedure of Section 10-2 guarantees input-output stability. This value of A will depend on the K_i you selected above.

10-3 Consider the circuit of Figure P10-3. The nonlinear capacitor has incremental capacitance

$$c(v) = \frac{k}{\sqrt{\phi - v}}$$

where K and ϕ are constants.

The charge-voltage relationship is such that $q(\phi)$ is zero. Let

$$e(t) = E + A \cos \omega_0 t \qquad \text{for all } t$$

Set up the problem in the form of Section 10-2. Find $\hat{h}(t)$, $h(t)$, $d(\bar{\mathbf{X}})$, and $\bar{\mathbf{X}}$ in terms of circuit constants.

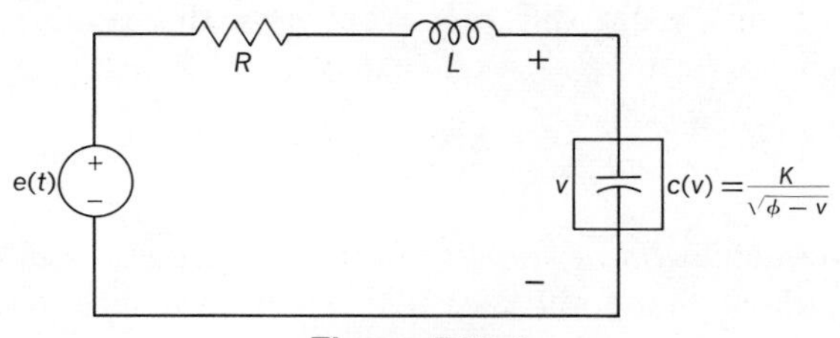

Figure P10-3

Bibliography

1. Bashkow, T. R., "The A Matrix, New Network Description," *IRE Trans.*, **CT-4** (1957), pp. 117–120.
2. Buck, R. C., *Advanced Calculus*, 2d Ed. New York: McGraw-Hill Book Company, Inc., 1965.
3. Carlin, H. J. and A. B. Giordano, *Network Theory*. Englewood Cliffs, N. J.: Prentice-Hall, Inc., 1964.
4. Cesari, L., *Asymptotic Behavior and Stability Theory in Ordinary Differential Equations*. New York: Academic Press, Inc., 1963.
5. Chua, L. O. and R. A. Rohrer, "On the Dynamic Equations of a Class of Nonlinear RLC Networks," *IEEE Trans.*, **CT-12**, No. 4 (1965), pp. 475–488.
6. Cooper, G. R. and C. D. McGillem, *Methods of Signal and System Analysis*. New York: Holt, Rinehart and Winston, Inc., 1967.
7. Cruz, J. B., Jr., and M. E. Van Valkenburg, *Introductory Signals and Circuits*. Waltham, Mass.: Blaisdell Publishing Company, 1967.
8. De Russo, P. M., R. J. Roy, and C. M. Close, *State Variables for Engineers*. New York: John Wiley & Sons, Inc., 1965.
9. Gantmacher, F. R., *The Theory of Matrices*, Vol. I. New York: Chelsea Publishing Company, 1960.
10. Gardner, M. and J. Barnes, *Transients in Linear Systems*. New York: John Wiley & Sons, Inc., 1942.
11. Golomb, M. and M. Shanks, *Elements of Ordinary Differential Equations*, 2d Ed. New York: McGraw-Hill Book Company, Inc., 1965.
12. Guillemin, E. A., *The Mathematics of Circuit Analysis*. New York: John Wiley & Sons, Inc., 1949.
13. Guillemin, E. A., *Synthesis of Passive Networks*. New York: John Wiley & Sons, Inc., 1957.
14. Guillemin, E. A., *Theory of Linear Physical Systems*. New York: John Wiley & Sons, Inc., 1963.
15. Hayashi, C., *Nonlinear Oscillations in Physical Systems*. New York: McGraw-Hill Book Company, Inc., 1964.
16. Hazony, D., *Elements of Network Synthesis*. New York: Reinhold Publishing Corporation, 1963.
17. Kuh, E. S. and R. A. Rohrer, "The State Variable Approach to Network Analysis," *Proc. IEEE*, **53**, No. 7 (1965), pp. 672–686.
18. Kuh, E. S. and R. A. Rohrer, *Theory of Linear Active Networks*, Holden-Day, 1967.

19. Liou, M. L., "A Novel Method of Evaluating Transient Response," *Proc. IEEE*, **54**, No. 1 (1966), pp. 20–22.
20. Lorens, S., *Flow Graphs*. New York: McGraw-Hill Book Company, Inc., 1964.
21. Paley, R. E. A. C. and N. Wiener, *Fourier Transforms in the Complex Domain*. New York: American Mathematical Society Colloquium Publications, 1934.
22. Schrawz, R. J. and B. Friedland, *Linear Systems*. New York: McGraw-Hill Book Company, Inc., 1965.
23. Slater, J. C. and N. H. Frank, *Mechanics*. New York: McGraw-Hill Book Company, Inc., 1947.
24. Sokolnikoff, I. S. and R. M. Redheffer, *Mathematics of Physics and Modern Engineering*. New York: McGraw-Hill Book Company, Inc., 1958.
25. Stern, T. E., *Theory of Nonlinear Networks and Systems*. Reading, Mass.: Addison-Wesley Publishing Company, Inc., 1965.
26. Tellegen, B. D. H., "A General Network Theorem with Applications," *Philips Res. Rep* **7** (1952), pp. 259–269.
27. Tricomi, F. G., *Integral Equations*. New York: Interscience Publishers, 1957.
28. Zadeh, L. A. and C. A. Desoer, *Linear System Theory*. New York: McGraw-Hill Book Company, Inc., 1963.
29. Zemanian, A. H., *Distribution Theory and Transform Analysis*. New York: McGraw-Hill Book Company, Inc., 1965.

Index

Index

A

A matrix, and system poles, 45–46
Algebraic properties of convolutions, 53–56
All pass factors, 154
Analog computer diagrams, 29–30
Approximate formulas, errors in, 95–100
Approximate response, *Z*-transforms of, 95
Asymptotic stability, 158

B

Bashkow, T. R., 15
Bessel's equation, 40
Block diagrams, change of state variables in, 69–72
  state variables for, 67–69
  steady-state analysis of, 65–67
  system repesentation, 64–72
Bode, H., 151
Bode procedure, 151, 155
Bounded stability, 158
Bounds on the response, 193–205

C

Capacitance loops, circuits with, 120–122
Capacitor loop, 20–24
Causality, 65*n*.
  of positive real systems, 150–158
Change of state variables, in block diagrams, 69–72
Characteristic polynomial, 9, 35
Circuits with capacitance loops or inductance cut sets, 120–122

Coates flow graph, 64
Component, definition of, 5
Constant of integration, 11
Controllability, 74–80
Convergence, of iterates, 199–200
Convolutional integral, general linear systems described by, 53–86
Convolutions, differentiation and integration of, 56–58
  with discontinuous functions, 61–63
  discrete, 91–93
  with infinite limits, 58–61
  properties of, 53–56
  transforms of, 138–139
Cramer's rule, 9, 10, 12

D

D'Alembert's principle, 14, 25, 111
Delta functions, 61–62, 80–81
Differential equations, and stability, 205–212
Differentiation, of convolutions, 56–58
Dirac delta function, 61–62, 80–81
Discrete convolutions, 91–93
Discrete systems, Z-transforms for, 102–105
Discrete time signal processing, 76–108
Discrete time signals, 87
Domain, definition of, 5

E

Eigenvalues, 15, 35, 45
Electric circuit diagram, 15–24
Element, definition of, 5
Energy function concepts, 169–185
Even and odd functions, 136–138

F

First-order difference equation, 100–102
Fixed coefficient systems, 34
Forced response, 12
Fourier cosine, 137
Fourier transform, 134–142
Frequency analysis, 42–45
Friedland, B., 5*n*.
Function, definition of, 5
Fundamental matrix, 36

G

Group delay, 154
Gyrator circuits, 23

H

High-order equations, 25–29
Hilbert transform, 139–142, 150, 152, 153, 155, 156, 157
Hold modulators, 88–89
Homogenous equations, 8–9, 34–35
Hurwitz polynomial, 145

I

Inductance cut set, circuits with, 120–122
Input-output relations, time-domain, for lumped systems, 87–91
Input-output stability, 158
Integral equations, and stability, 205–212
Integration, of convolutions, 56–58
Interconnected systems, Z-transforms of (*table*), 98
Inverse Z-transforms, 93
Iterates, 187
  convergence of, 189–192, 199–200
  first, 194–196
  sequence of, 187–189
  upper bounds for successive, 197–198
Iteration, 187
  of modified equation, 196–197

of nonlinear systems problems, 186–213

J

Jacobian matrix, 181
Jordan form, 77*n.*

K

Kirchhoff's current law (*KCL*), 7, 14, 16, 24, 111
Kirchhoff's voltage law (*KVL*), 14, 16, 24, 40, 111

L

Laplace transforms, 7, 42–43, 50
Laplace-Fourier transforms, 134–142
Leibniz formula, 57
Limit cycle, 176
Linear discrete time systems, 100–106
Linear system, definition of, 6
Linearity, 72–74
Lipschitz condition, 189*n.*
Logarithmic decrement, 154
Loop of capacitors, 20–24
Lossless electric circuits, 118–120
Lossless systems, 113–116
Lumped systems, characterized by positive real matrices, 143–165
  definition of, 5
  time-domain input-output relations for, 87–91
  Z-transform analysis of, 94–100
Lyapunov stability, 178, 181, 205, 208

M

Magnitude, asymptotic limits on, 156–158
  and phase relations, 152–155
Mapping, definition of, 5
Mathieu's equation, 40
Matrices, 36–39, 45–50
Matrix difference equations, *Z*-transform of, 105–106
Mechanical circuit diagram, 24–25
Minimum phase transfer function, 153
Modulators, 88–91
  hold, 88–89
  pulse amplitude, 89–90
  pulse width, 90–91
  sample, 88–89
Mutual inductance, 22

N

Natural frequencies of the system, 35
Natural response, 12

O

Observability, 74–80

P

Paley-Wiener criterion, 157, 195
Passivity, for positive real systems, 160–162
  for positive-semidefinite systems, 123–124
PD (*see* Positive definite system)
Phase, asymptotic limits on, 156–158
  and magnitude relations, 152–155
Phase plane, 176
Physical, definition of, 5
Poles and zeros of positive real matrix elements, 143–150
Positive definite system (*PD*), 113
Positive definiteness, tests for, 113–116
Positive real matrix (*PR*), 124
  causality of systems of, 150–158
  poles and zeros of elements of, 143–150
  properties of lumped systems characterized by, 143–165
  transfer function, 124–125
Positive real 1-ports, 144–148

Positive real systems, passivity for, 160–162
stability for, 158–159
Positive real 2-ports, 148–150
Positive-semidefinite system (*PSD*), 113
passivity for, 123–124
PR (*see* Positive real matrix)
PSD (*see* Positive-semidefinite system)
Pulse amplitude modulators, 89–90
Pulse width modulators, 90–91

Q

Quadrantal symmetry, 151

R

Range, definition of, 5
Real and imaginary parts, relations between, 150–158
Resistive systems, 113–116
Response of the model, 3

S

Sample modulators, 88–89
Schwartz, R. J., 5*n.*
Schwartz's inequality, 190
Second-order systems, 174–177
Signal, definition of, 5
Signal flow graph, 64
Sine transform, 137
Sine wave, 43
Stability, 177–182
Lyapunov, 178, 181, 205, 208
for positive real systems, 158–159
using both integral and differential equations, 205–212
State at t_0, 72–74, 178, 187
as an additional input, 80
definition of, 6
State space, 38, 174
State transition matrix, 37–39, 46–50
State variable concepts, 169–185
State variables, 38
for block diagrams, 67–69
State vector, 38
Steady-state analysis, of block diagrams, 65–67
Steady-state response, 44
Steady-state transfer functions, 44–45
System, definition of, 4, 5
System diagram, 14
System poles, and *A* matrix, 45–46
Systems with positive-semidefinite energy functions, 111–133
lossless systems, 116–122
resistive systems, 113–116

T

Taylor's theorem, 181
Tellegen's theorem, 113
Terminology, 4–6
Time-domain input-output relations for lumped systems, 87–91
Time-invariance, 72–74
Time-invariant system, definition of, 6
Time-invariant systems, 39–41
Total response, 72–83
Trajectories in the phase plane, 174–177
Transfer functions, 44–45, 124–125, 150–156
Transforms, 7, 42–43, 50, 91–93, 94–100, 134–142
Two-sided transforms, 134–136

V

V-functions, 180
Variation of parameters, 4, 7–8, 9–12, 13
solution of normal form equations by, 34–52

Volterra theory, 171

Z

Z-transforms, 91–93
  analysis of lumped systems, 94–100
  of the approximate response, 95
  for discrete systems, 102–105
  of interconnected systems (*table*), 98
  inverse, 93
  of matrix difference equations, 105–106
Zero-input response, 12–44
Zero-state plus zero-input, 72–83
Zero-state response, 12